Advances in Solid-State Laser Technology and Applications

Advances in Solid-State Laser Technology and Applications

Guest Editor

Zhen Qiao

MDPI

Basel • Beijing • Wuhan • Barcelona • Belgrade • Novi Sad • Cluj • Manchester

Guest Editor
Zhen Qiao
School of Optical-Electrical
and Computer Engineering
University of Shanghai for
Science and Technology
Shanghai
China

Editorial Office
MDPI AG
Grosspeteranlage 5
4052 Basel, Switzerland

This is a reprint of the Special Issue, published open access by the journal *Photonics* (ISSN 2304-6732), freely accessible at: https://www.mdpi.com/journal/photonics/special_issues/L792032X4V.

For citation purposes, cite each article independently as indicated on the article page online and as indicated below:

Lastname, A.A.; Lastname, B.B. Article Title. *Journal Name* **Year**, *Volume Number*, Page Range.

ISBN 978-3-7258-8147-5 (Hbk)
ISBN 978-3-7258-8148-2 (PDF)
https://doi.org/10.3390/books978-3-7258-8148-2

Contents

About the Editor . . . . **vii**

Mengyao Cheng, Hua Wang, Wenlong Tian, Yizhou Liu and Jiangfeng Zhu
Research on Nanosecond High-Pulse-Energy Regenerative Amplifier with Adjustable Pulse Duration and Third Harmonic Generation
Reprinted from: *Photonics* **2025**, *12*, 353, https://doi.org/10.3390/photonics12040353 . . . . **1**

Xian-An Dou, Linchan Li, Chang Liang, Haiping Xu, Qing Ye, Hui Kong, et al.
Nonlinear-Optical-Loop-Mirror-Based Mode-Locked Fiber Laser Sensor for Low-Temperature Measurement
Reprinted from: *Photonics* **2025**, *12*, 507, https://doi.org/10.3390/photonics12050507 . . . . **11**

Shuzhen Nie, Jinglan Lin, Tianzhuo Zhao and Xiaolong Liu
Exploring the Impact of Inlet Velocity Distribution on the Thermal Performance of a Laser Rod in a Diode Side-Pumped Amplifier
Reprinted from: *Photonics* **2025**, *12*, 603, https://doi.org/10.3390/photonics12060603 . . . . **19**

Songtao Li, Shengxi Zheng, Bowen Zheng, Yong Wei, Yongchang Zhang, Yanmin Duan, et al.
Visible Triple-Wavelength Switchable Emission Generated in Passively Q-Switched Nd:YVO_4 Self-Raman Laser
Reprinted from: *Photonics* **2025**, *12*, 669, https://doi.org/10.3390/photonics12070669 . . . . **31**

Bosong Yu, Siying Wang, Aimin Wang, Yizhou Liu and Lishuang Feng
Miniaturized Multicolor Femtosecond Laser Based on Quartz-Encapsulated Nonlinear Frequency Conversion
Reprinted from: *Photonics* **2025**, *12*, 836, https://doi.org/10.3390/photonics12090836 . . . . **41**

Qianyu Yao, Lilin Luo, Yue Cai, Yongguo Zheng and Xinhai Zhang
Switchable Dual-Comb Fiber Laser Based on Sagnac Loop for Wavelength-Multiplexing and Polarization-Multiplexing
Reprinted from: *Photonics* **2025**, *12*, 961, https://doi.org/10.3390/photonics12100961 . . . . **52**

John J. Degnan
Extending the ICESAT-2 ATLAS Lidar Capabilities to Other Planets Within Our Solar System
Reprinted from: *Photonics* **2025**, *12*, 1048, https://doi.org/10.3390/photonics12111048 . . . . . . **67**

Shuaichen Li, Yiyang Zhang, Ying Li, Linge Mao, Pengfan Zhao and Zhen Qiao
Tunable-Charge Optical Vortices Through Edge Diffraction of a High-Order Hermit-Gaussian Mode Laser
Reprinted from: *Photonics* **2025**, *12*, 1076, https://doi.org/10.3390/photonics12111076 . . . . . . **83**

Xiaochuan Zheng, Yanhua Lu, Xuguang Zhang, Xingwang Luo, Junzhi Ye, Peng Huang, et al.
154-W, Single-Frequency, Two-Stage Innoslab Amplifier at 1319 nm
Reprinted from: *Photonics* **2026**, *13*, 449, https://doi.org/10.3390/photonics13050449 . . . . **95**

About the Editor

Zhen Qiao

Zhen Qiao is an Associate Professor in Optical Engineering at the School of Optical-Electrical and Computer Engineering, University of Shanghai for Science and Technology (USST), Shanghai, China. His research focuses on the spatial mode manipulation of lasers, vortex laser intracavity control technologies, and structured light field generation and regulation, with particular emphasis on orbital angular momentum lasers and microcavity lasers. He actively explores applications of intracavity structured light in optical information processing, biosensing, and biomedical imaging. He has authored more than 50 SCI-indexed papers in prestigious journals including *Nature Communications, Optica, Advanced Photonics*, and *Laser Photonics Reviews*, and holds four authorized invention patents. His research has been featured by scientific media outlets such as Laser Focus World and Phys. Org. He currently leads a project funded by the National Natural Science Foundation of the China Youth Program and is supported by the Shanghai Overseas High-Level Talent Introduction Program. He also serves as a Guest Editor for *Photonics*.

photonics
MDPI

Article

Research on Nanosecond High-Pulse-Energy Regenerative Amplifier with Adjustable Pulse Duration and Third Harmonic Generation

Mengyao Cheng [1], Hua Wang [2], Wenlong Tian [2], Yizhou Liu [3] and Jiangfeng Zhu [2,*]

1 School of Artificial Intelligence, Shaanxi Institute of Technology, Xi'an 710300, China; mengyaocheng@126.com
2 School of Optoelectronic Engineering, Xidian University, Xi'an 710071, China; wanghua@stu.xidian.edu.cn (H.W.); wltian@xidian.edu.cn (W.T.)
3 School of Information Science and Engineering and Shandong Provincial Key Laboratory of Laser Technology and Application, Shandong University, Qingdao 266237, China; yizhou.liu@sdu.edu.cn
* Correspondence: jfzhu@xidian.edu.cn

Abstract: We reported on a nanosecond regenerative amplified laser with a repetition rate of 1 kHz by employing laser diodes (LDs) with distinct wavelengths as both the seed laser and the pump source and utilizing Nd:YAG as the gain medium. The single-pulse energy was 1.58 mJ and the pulse duration was adjustable, ranging from 1 to 5 ns. Combining two oppositely oriented BBO crystals for second harmonic generation (SHG) and an LBO crystal for third harmonic generation (THG), a 355 nm laser with a single-pulse energy of 257 μJ was attained, corresponding to a THG efficiency of 16.2%.

Keywords: nonlinear optics; frequency conversion; diode-pumped lasers; amplifiers

1. Introduction

Compared to the direct generation of nanosecond infrared lasers from Nd-doped crystals, the high-energy nanosecond ultraviolet laser produced through amplification and third harmonic generation (THG) exhibits superior characteristics, including a higher single-pulse energy, elevated photon energy, and a smaller focal spot size. This laser has been widely applied in research, medical, and industrial fields. Its notable applications include fluorescence lifetime imaging [1], laser-induced breakdown spectroscopy [2], laser surface texturing [3], laser annealing [4–6], the laser processing of ceramics [7–10], and terahertz generation [11]. A particularly compelling application is in the field of radio-photoluminescence (RPL) [12], where there is a pronounced demand for a high-energy nanosecond ultraviolet laser with adjustable pulse duration capabilities.

Traditional nanosecond laser amplifiers typically utilize either Q-switched solid-state lasers [13] or fiber lasers [14,15] as their seed source. Q-switching technology can be categorized into two types: active Q-switching [16] and passive Q-switching [17,18]. High-energy nanosecond lasers often employ passively Q-switched lasers with pulse durations below 10 ns as their seed source. However, these systems lack the ability to adjust their pulse duration flexibly. With the progress in LDs and their driving technologies, using LDs with a center wavelength of 1064 nm and pulse durations below 10 ns as seed sources has become a viable alternative. By this method, the arbitrary adjustment of the regenerative amplifier's output laser pulse duration can be achieved. In addition, using Q-switched lasers as seed sources requires additional components, such as a pump source, attenuation modules, and pulse-selection systems to control pulse energy and repetition rate. This not only increases the overall cost but also compromises system stability. In contrast, nanosecond high-pulse-energy regenerative amplifiers utilizing LDs as seed sources offer significant advantages, including a simple structure, low cost, and continuous pulse duration adjustability. These features make them the optimal choice for achieving high-pulse-energy, pulse-duration-adjustable nanosecond lasers.

Laser amplifiers are broadly classified into two categories: traveling-wave amplifiers and regenerative amplifiers. For nanosecond lasers with repetition rates exceeding 10 kHz, multi-stage traveling-wave amplification can mitigate period-doubling bifurcation [19], achieve higher average power, and generate burst-mode pulse trains. However, for applications requiring lower repetition rates and higher single-pulse energy, particularly when seeding with an LD, regenerative amplifiers are more suitable to amplify the seed to higher energy levels while maintaining excellent beam quality and enabling a more compact system design.

In experiments involving ultraviolet nanosecond laser generation via THG, a two-step process involving initial frequency doubling followed by frequency mixing is often preferred due to the low efficiency of direct THG. In this process, the commonly used nonlinear crystals include LBO [20], BBO [21], and KDP [21], whose physical and optical properties are summarized in Table 1.

Table 1. Parameters of nonlinear crystals.

Crystals	Transmission Range /nm	Damage Threshold [22] /GW·cm^{-2}	$d_{eff\,1}$ [1] /pm·V^{-1}	$d_{eff\,2}$ [2] /pm·V^{-1}	Walk-Off Angle [1] /mrad	Walk-Off Angle [2] /mrad
LBO	160–2600	36.3	0.83 (xy)	0.53 (yz)	7.06 (xy)	9.30 (yz)
BBO	185–2600	18.27	2.01	2.02	55.86	72.33
KDP	177–1700	20	0.26	0.32	28.06	30.12

[1] Under the condition of 1064 nm frequency doubling output at 532 nm. [2] Under the condition of 1064 nm and 532 nm sum frequency output at 355 nm.

From the data presented in the table, BBO crystals can be seen to exhibit the highest nonlinear coefficient, while LBO crystals have the highest damage threshold. Therefore, in nanosecond-pulse THG experiments at lower peak power, it is advantageous to use BBO crystals for SHG and LBO crystals for THG to improve the efficiency and reduce the risk of crystal damage. Additionally, given the pronounced walk-off angle of BBO crystals, using two oppositely oriented BBO crystals for SHG can effectively compensate for the spatial walk-off effect.

For nanosecond ultraviolet lasers, several promising results have demonstrated high conversion efficiency [23]. However, these lasers typically rely on a conventional Q-switched laser, which inherently lack the capability for broad-range, arbitrary-pulse-width tuning. Some studies have used LDs as the seed source for all-fiber regenerative amplifiers [24], but such lasers struggle to achieve a high pulse energy. In this work, we employ an LD as the seed source for an all-solid-state laser regenerative amplifier and THG to produce a high-energy, nanosecond ultraviolet laser with a fully adjustable pulse width.

In summary, a commercial LD with a center wavelength of 1064.3 nm, a peak power of 30 W, and a repetition rate of 1 kHz was used as the seed source for the amplifier. The seed laser's pulse duration was tunable from 1 to 5 ns. For the regenerative amplifier, a fiber-coupled LD served as the pump source, with an Nd:YAG crystal as the gain medium. The regenerative amplification process increased the pulse energy to 1.58 mJ while keeping the pulse duration at 1–5 ns. The amplified laser pulses were focused in order to increase power density, enabling ultraviolet laser generation via two BBO crystals and one LBO crystal. As a result, the system produced a 355 nm laser of 1 ns pulse duration with a single-pulse energy of 257 μJ and a THG efficiency of 16.2%.

2. Experimental Setup

The optical structure of the experimental setup is shown in Figure 1. The experimental setup comprised a seed source, a regenerative amplifier, a THG unit, and an electronic control system.

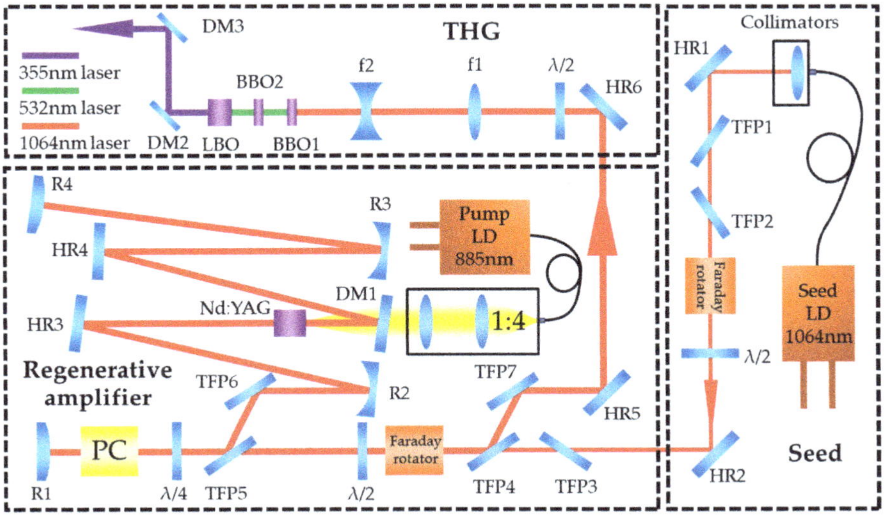

Figure 1. Optical structure of experimental device. (HR, high-reflectivity mirror; DM, dichroic mirror; TFP, thin-film polarizer).

The seed source was a commercial LD with a maximum peak power of 30 W and an adjustable output pulse duration ranging from 1 to 10 ns. By adjusting its operating temperature, the central wavelength could be set at 1064.3 nm, which matched the emission peak of the gain medium in the regenerative amplifier. The repetition rate of the seed LD was adjustable within the range of 0.1–1 kHz, and the seed laser was coupled through a single-mode polarization-maintaining fiber. After using a collimator, the seed laser became horizontally polarized and was injected into two 45° high-reflectivity mirrors (HR1 and HR2). By adjusting the angles of HR1 and HR2, the seed laser could be precisely directed into the regenerative amplifier. Additionally, an isolator composed of two thin-film polarizers (TFP1 and TFP2), a Faraday rotator, and a half-wave plate (λ/2) was inserted between HR1 and HR2 to prevent laser feedback in the regenerative amplifier from damaging the seed source. After the first isolator, the seed laser passed through a second isolator consisting of TFP3, TFP4, a Faraday rotator, and a λ/2 plate, maintaining its horizontal polarization. Subsequently, the laser was transmitted through TFP5 and entered the regenerative amplifier. The regenerative amplifier utilized a fiber-coupled LD with a central wavelength of 885 nm as its pump source. This pump source had a maximum output power of 70 W and used a coupling fiber with a 200 μm core diameter. The pump beam delivered from the fiber was focused into the laser gain medium by a coupling system with magnification of 1 and a dichroic mirror (DM1) with both anti-reflection coating at 885 nm and high-reflectance coating at 1064 nm. The focused beam waist in the gain medium was about 800 μm. To mitigate excessive thermal effects in the gain medium that could degrade amplification efficiency, the pump source was operated in a pulsed mode synchronized with the regenerative amplifier, featuring a 230 μs pulse duration. The regenerative amplifier employed a Nd:YAG crystal (16 mm length, 4×4 mm^2 cross-section) with a 0.15-at.% Nd^{3+} doping concentration as the gain medium. For thermal management, the crystal was encapsulated in indium foil and secured in a gold-plated copper heat sink with water-cooling. To suppress Fabry–Perot etalon effects, all of the crystal's surfaces were deposited with anti-reflection coatings optimized for both 885 nm and 1064 nm, and the crystal was fabricated with a 2° wedge angle. The regenerative amplifier had a total cavity length of 1236 mm, and the symmetric confocal cavity was formed by two concave mirrors, R2 and R3, each with a curvature radius of 900 mm. The end mirrors on both sides of the amplifier were convex mirrors, R1 and R4, with curvature radii of −600 mm. The distances from R2 to the crystal and from R3 to the crystal were both 350 mm. Similarly, the separation between R1 and R2, and between R3 and R4, was 260 mm. To achieve a compact optical layout, plane mirrors HR3 and HR4 were inserted into the cavity to fold the optical path. Using ABCD matrix calculations, the laser beam diameter at the crystal was set to be approximately 1 mm, ensuring excellent mode matching with the 800 μm pump beam. To enable injection and cavity dumping in the regenerative amplifier, a Pockels cell (PC) with a 4 mm aperture and a $\lambda/4$ waveplate were inserted between TFP5 and R1. Additionally, TFP6 was placed between TFP5 and R2 to optimize the optical path. After amplification, the laser exited through TFP5 and passed through a $\lambda/2$ waveplate and a polarizer, where its polarization direction was rotated to the vertical direction. The beam was then reflected by TFP7 and a 45° high-reflectivity mirror (HR5) before entering the THG unit.

In the THG unit, the amplified laser was first reflected by a 45° high-reflectivity mirror (HR6). A $\lambda/2$ waveplate ($\lambda/2$ 3) was inserted after HR6, which allowed us to adjust the incident laser's polarization to achieve a higher conversion efficiency. Due to the large beam diameter and divergence of the amplified laser, a pair of lenses (f1 and f2) were used to reduce the beam size and collimate it after entering the THG unit. This improved the power density and enhanced the conversion efficiency. The focal lengths of f1 and f2 were 150 mm and −50 mm, respectively, and the distance between f1 and f2 was approximately 100 mm. The frequency doubling utilized two BBO crystals placed in opposite directions to compensate for the walk-off effect, as illustrated in Figure 2. Both BBO crystals had a length of 6 mm and a cross-sectional area of 4 × 4 mm^2, with cutting angles of θ = 22.8° and φ = 0°. The crystals were coated on both sides with dielectric films highly transparent to both 1064 nm and 532 nm lasers. For sum frequency generation, a single LBO crystal was used, which was 10 mm in length with a cross-sectional area of 4 × 4 mm^2 and cutting angles of θ = 44.2° and φ = 90°. One side of the LBO crystal was coated with a dielectric film highly transparent to both 1064 nm and 532 nm lasers, and the other side was coated with a dielectric film transparent to 1064 nm, 532 nm, and 355 nm lasers. To achieve precise temperature control of the nonlinear crystals, both crystals were encapsulated in indium foil and secured in a gold-plated copper heat sink with thermoelectric coolers (TECs). The other side of each TEC was cooled by a water-cooled housing. The two BBO crystals shared one set of heat sinks and TECs, while the LBO crystal used another set, allowing for independent temperature controlling. The temperature control range was from 10 °C to 50 °C. After the nonlinear crystals, the laser was reflected by two 45° dichroic mirrors (DM2 and DM3) and then exited through a window plate. Both sides of DM2 and DM3 were coated with dielectric films that highly reflect 355 nm lasers and transmit both 532 nm and 1064 nm lasers, ensuring a pure 355 nm laser output. The window plate was an uncoated fused silica glass plate, positioned at the Brewster angle to achieve high transmission efficiency. Additionally, the window plate featured a sealed design to prevent the contamination of the nonlinear crystals by external air and dust.

To enhance the stability of the laser system, the regenerative amplifier and the THG unit were placed inside a sealed aluminum housing. The housing had a footprint of 570 × 375 mm^2 and a height of 150 mm. To prevent damage to the nonlinear crystals, the regenerative amplifier and THG stages were isolated by the housing, while the THG cavity was equipped with replaceable desiccant and specially engineered low-outgassing cables. Next, the seed laser entered the housing through a sidewall-mounted collimator, was amplified by the regenerative amplifier, and was frequency-tripled by the THG unit before exiting through a window plate. The aluminum housing was cooled by circulating water maintained at 24 °C to ensure stable thermal conditions. Heat-generating components, such as the gain medium of the regenerative amplifier, the pump LD, and the TECs for the temperature control of the nonlinear crystals, were all cooled through the housing.

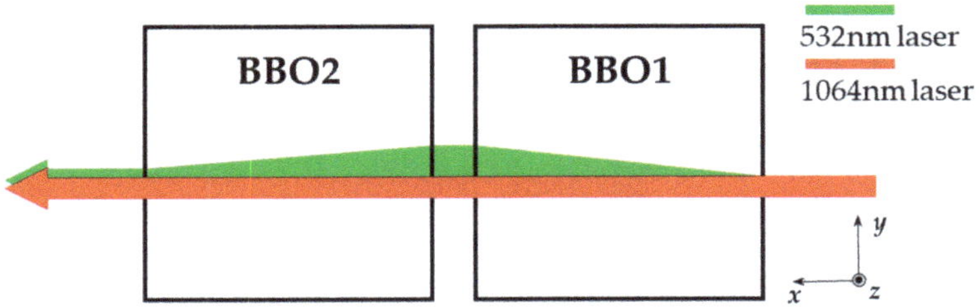

Figure 2. Schematic diagram of space walk-off and compensation for double BBO crystals.

3. Experimental Results

In the laser system, the pulse duration of the seed source LD could be adjusted within the range of 1–10 ns. However, due to the regenerative amplifier cavity length being only 1236 mm, it could only support a maximum pulse duration of 5 ns. Initially, the pulse duration of the seed source was set to 1 ns, and the peak power of the seed pulse output by the seed source reached 30 W, with a repetition frequency of 1 kHz. With the seed source LD's peak power remaining constant, the single-pulse energy scaled proportionally with pulse width. Specifically, at a seed pulse width of 1 ns, the minimum single-pulse energy was 30 nJ, while at a 5 ns pulse width, it increased to 150 nJ. Simultaneously, the seed source also output a synchronization signal identical to the seed pulse to synchronize with the regenerative amplifier.

During the process of adjusting the regenerative amplifier, we first used the seed laser to adjust all the cavity mirrors until the beam could retro-reflect precisely at both end mirrors (R1 and R4 in Figure 1). And the Nd:YAG crystal was intentionally tilted slightly in the horizontal plane to prevent parasitic lasing. Finally, after blocking the seed laser, we carefully adjusted the $\lambda/4$ waveplate to achieve quasi-continuous output. At this stage, the pump source operated at a repetition frequency of 1 kHz with a pulse duration of 230 μs, corresponding to a duty cycle of 23%. By meticulously optimizing parameters such as the end mirror angle and the pump coupling distance, and measuring the output laser power with a power meter, we achieved a quasi-continuous output power of 2.17 W at a maximum current of 7.5 A in the pump source LD. Subsequently, we activated the Pockels cell and fine-tuned the $\lambda/4$ waveplate. Through the careful adjustment of the Pockels cell voltage, the delay between the Pockels cell and the pump source, the Pockels cell gate width, and the $\lambda/4$ waveplate angle, a Q-switched output with a maximum output power of 1.63 W at 7.5 A was achieved. Finally, the seed source was injected into the regenerative amplifier. By precisely adjusting the injection direction of the seed source, the delay between the seed source and the Pockels cell, and the Pockels cell gate width, we achieved a regenerative amplification output power of 1.58 W at the same current of 7.5 A. The output powers under quasi-continuous, Q-switched, and regenerative amplification conditions are summarized in the Table 2:

Table 2. Output power under quasi-continuous, Q-switched, and regenerative amplified conditions.

Current of Pump Source LD/A	Quasi-Continuous Output Power/W	Q-Switched Power/W	Regenerative Amplified Output Power/W
4.5	0.35	0	0
5	0.62	0	0
5.5	0.89	0	0
6	1.20	0.34	0.15
6.5	1.54	1.02	0.78
7	1.87	1.46	1.33
7.5 [1]	2.17	1.63	1.58

[1] The peak power of the pump source at a current of 7.5 A was 50 W.

To prevent potential damage to the nonlinear crystals in the THG unit caused by excessive power, we ultimately conducted subsequent experiments at a pump current of 7.5 A. Under these conditions, the peak power of the pump source was approximately 50 W, the gate width of the Pockels cell was 341 ns, and the output power of the regenerative amplifier was 1.58 W, with a repetition rate of 1 kHz, corresponding to a single-pulse energy of 1.58 mJ. A photodiode and an oscilloscope were used to measure the regenerative amplification pulse buildup sequence and the output laser waveform by detecting the leakage laser from the cavity mirror and the output laser of the regenerative amplifier, respectively. The results are illustrated in Figure 3 below:

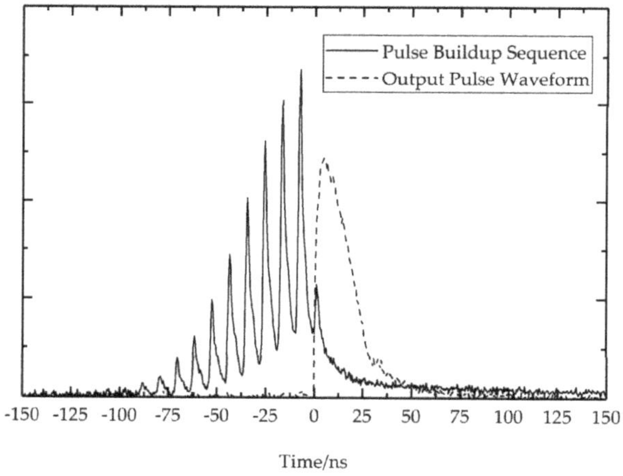

Figure 3. Pulse buildup waveform and output pulse waveform of regenerative amplifier.

Due to the limited bandwidth of the photodiode and oscilloscope we employed, we were unable to accurately measure the pulse duration of the output laser from the regenerative amplifier. We also measured the output power and pulse buildup waveforms of the regenerative amplifier at different pulse durations. It was found that within the seed source pulse duration range of 1–5 ns, the output power and pulse buildup waveforms of the regenerative amplifier did not exhibit significant changes. Additionally, we used a CCD camera to measure the output beam profile of the regenerative amplifier. The measurements revealed that the long-axis diameter of the output beam was 2.26 mm and the short-axis diameter was 2.16 mm, corresponding to a beam roundness of 96%. An image of the output beam profile is shown in Figure 4:

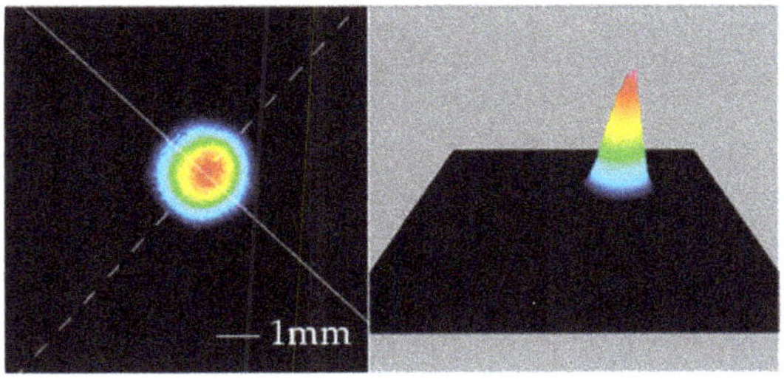

Figure 4. Output laser spot of regenerative amplifier.

In the THG process, due to the large and divergent beam output from the regenerative amplifier, a pair of lenses with focal lengths of f = 150 mm and f = −50 mm were employed to reduce and collimate the beam. Subsequently, by meticulously adjusting parameters such as the incident direction and polarization direction of the fundamental laser, the spacing between the lenses, and the angles and temperatures of the nonlinear crystals, an ultraviolet laser with a single-pulse energy of 257 μJ was achieved under the conditions of a 1.58 mJ single-pulse energy, 1 ns pulse duration, and 1 kHz repetition rate of the infrared laser. The corresponding THG efficiency was 16.2%. At this point, the temperatures of the BBO crystals and LBO crystal were 24.0 °C and 45.8 °C, respectively. During the adjustment process, it was also found that the temperature of the LBO crystal had a significant impact on the efficiency, whereas the temperature of the BBO crystals had a relatively minor effect. This is because the LBO crystal had a higher thermal birefringence coefficient compared to the BBO crystal [21], along with a smaller nonlinear optical coefficient, making its phase-matching conditions more sensitive to temperature variations. The spectrum of the ultraviolet laser, measured using a spectrometer, is shown in Figure 5:

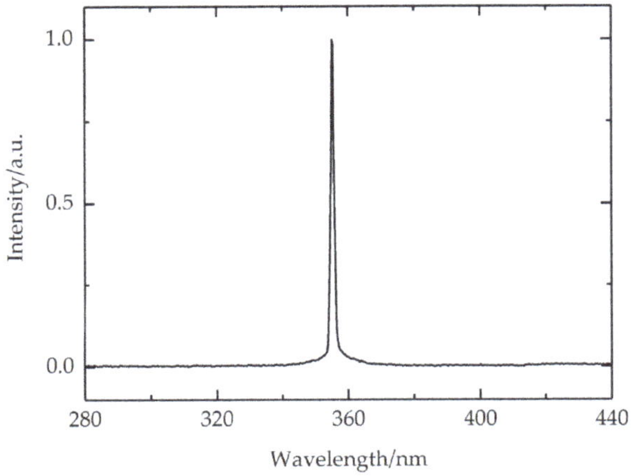

Figure 5. Spectrum of ultraviolet laser.

We also tested the THG efficiency at different pulse durations, with the highest efficiency of 16.2% being at 1 ns and the lowest of 4.68% at 5 ns. This is because the peak power density of the laser is higher at shorter pulse durations, leading to a higher THG efficiency. The variation in THG efficiency at different pulse durations also indirectly demonstrated that using an LD as a seed source enables the adjustment of the output laser pulse duration from the amplifier. The curve of the ultraviolet laser output power at different pulse durations is shown in Figure 6:

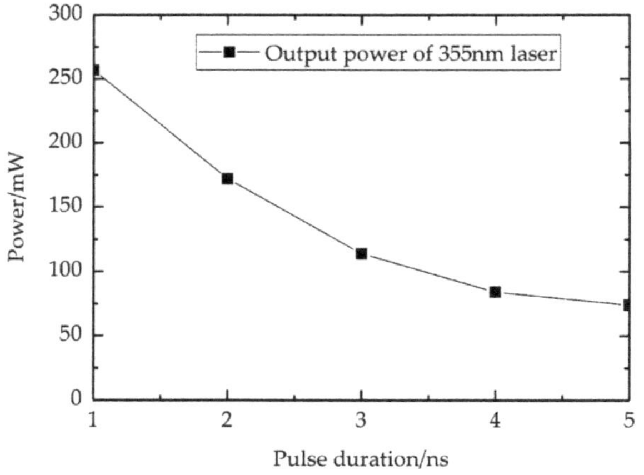

Figure 6. Output power of ultraviolet laser at different pulse durations.

Finally, after sealing the housing, we measured the power stability of the laser over a period of one hour. The root mean square (RMS) error of the average power was less than 1%, indicating that the laser exhibits excellent overall stability and can operate reliably in the long term as a commercial laser. The power stability curve is shown in Figure 7:

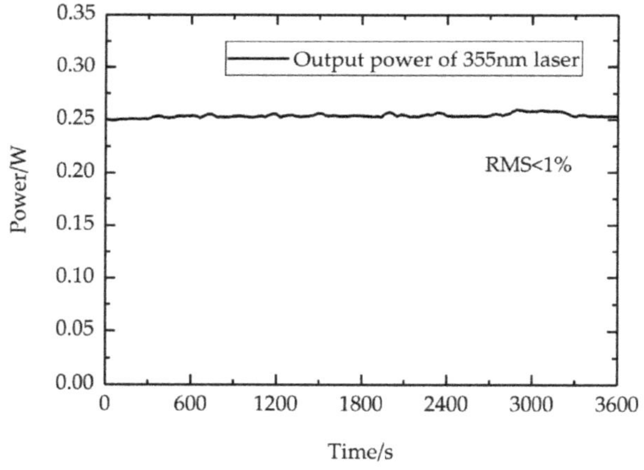

Figure 7. Stability of ultraviolet laser output power.

4. Summary and Outlook

In summary, using LDs as the seed and pump sources, we demonstrated a Nd:YAG nanosecond regenerative amplifier with a repetition rate of 1 kHz, a single-pulse energy of 1.58 mJ, and an adjustable pulse duration ranging from 1 to 5 ns. By employing two counter-positioned BBO crystals and one LBO crystal for THG, we obtained a maximum output of 257 μJ at 355 nm with a pulse duration of 1 ns, corresponding to a THG efficiency of 16.2%, which is notably high for nanosecond ultraviolet lasers [23,25]. The entire laser system was enclosed in a sealed aluminum box measuring $570 \times 375 \times 150$ mm^3, with a power stability of less than 1% over one hour. This nanosecond 355 nm laser with tunable pulse duration and high-power energy can be widely used in fluorescence lifetime imaging, laser-induced breakdown spectroscopy, laser surface texturing, laser annealing, the laser processing of ceramics, and terahertz generation. In future work, we plan to increase the infrared laser output power by adding a traveling-wave amplification stage, and further improve the THG efficiency by optimizing the nonlinear crystal parameters.

Author Contributions: Conceptualization, M.C. and J.Z.; methodology, W.T.; software, M.C.; validation, M.C. and H.W.; formal analysis, W.T.; investigation, M.C.; resources, J.Z.; data curation, M.C.; writing—original draft preparation, M.C.; writing—review and editing, W.T.; visualization, W.T.; supervision, J.Z.; project administration, J.Z.; funding acquisition, Y.L. All authors have read and agreed to the published version of the manuscript.

Funding: This research was funded by the Shaanxi Provincial Department of Education Key Scientific Research Program Projects (24JR017) and the Annual Scientific Research Plan of Shaanxi Institute of Technology (Pt24-01).

Institutional Review Board Statement: Not applicable.

Informed Consent Statement: Not applicable.

Data Availability Statement: The data presented in this study are available on request from the corresponding author. The data are not publicly available due to the fact that the data also form part of an ongoing study.

Conflicts of Interest: The funders had no role in the design of the study; in the collection, analyses, or interpretation of data; in the writing of the manuscript; or in the decision to publish the results.

References

1. Awasthi, K.; Nakabayashi, T.; Ohta, N. Application of Nanosecond Pulsed Electric Fields into HeLa Cells Expressing Enhanced Green Fluorescent Protein and Fluorescence Lifetime Microscopy. *J. Phys. Chem. B* **2012**, *116*, 11159–11165. [CrossRef] [PubMed]
2. Zhang, D.; Zhang, Z.; Zhang, M.; Feng, Z.; Gu, T.; Shen, K.; Li, S.; Hou, J.; Zhu, J. Portable nanosecond laser for handheld laser-induced breakdown spectroscopy instruments. *Opt. Eng.* **2023**, *62*, 036102. [CrossRef]
3. Bakhtiyari, A.N.; Yadav, A.; Wu, Y.; Zheng, H. Enhancing wettability of sapphire (Al_2O_3) substrate through laser surface texturing. *Mater. Lett.* **2024**, *355*, 135506. [CrossRef]
4. Park, S.Y.; Choi, Y.; Seo, Y.H.; Kim, H.; Lee, D.H.; Truong, P.L.; Jeon, Y.; Yoo, H.; Kwon, S.J.; Lee, D.; et al. 355 nm Nanosecond Ultraviolet Pulsed Laser Annealing Effects on Amorphous In-Ga-ZnO Thin Film Transistors. *Micromachines* **2024**, *15*, 103. [CrossRef]
5. Wang, H.; Zhang, F.; Wang, C.; Duan, J. Broadband antireflective subwavelength structures for infrared detection. *Opt. Laser Technol.* **2022**, *149*, 107931. [CrossRef]
6. Ding, Y.; Liu, L.; Wang, C.; Li, C.; Lin, N.; Niu, S.; Han, Z.; Duan, J. Bioinspired Near-Full Transmittance MgF_2 Window for Infrared Detection in Extremely Complex Environments. *ACS Appl. Mater. Interfaces* **2023**, *15*, 30985–30997. [CrossRef]
7. Jia, X.; Lin, J.; Li, Z.; Wang, C.; Li, K.; Wang, C.; Duan, J. Continuous wave laser ablation of alumina ceramics under long focusing condition. *J. Manuf. Process.* **2025**, *134*, 530–546. [CrossRef]
8. Li, Z.; Lin, J.; Jia, X.; Li, X.; Li, K.; Wang, C.; Sun, K.; Ma, Z.; Duan, J. High efficiency femtosecond laser ablation of alumina ceramics under the filament induced plasma shock wave. *Ceram. Int.* **2024**, *50*, 47472–47484. [CrossRef]
9. Jia, X.; Luo, J.; Guo, C.; Li, Z.; Ma, Z.; Xiang, Y.; Yi, Z.; Li, K.; Wang, C.; Li, X.; et al. High-energy continuous wave laser ablation of alumina ceramic. *J. Mater. Res. Technol.* **2023**, *27*, 5389–5403. [CrossRef]
10. Jia, X.; Li, Z.; Wang, C.; Li, K.; Zhang, L.; Ji'an, D. Study of the dynamics of material removal processes in combined pulse laser drilling of alumina ceramic. *Opt. Laser Technol.* **2023**, *160*, 109053. [CrossRef]
11. Hayashi, S.; Shibuya, T.; Sakai, H.; Taira, T.; Otani, C.; Ogawa, Y.; Kawase, K. Tunability enhancement of a terahertz-wave parametric generator pumped by a microchip Nd: YAG laser. *Appl. Opt.* **2009**, *48*, 2899–2902. [CrossRef] [PubMed]
12. Maki, D.; Sato, F.; Kato, Y.; Yamamoto, T.; Iida, T. Improvement of a dose reading system for radiophotoluminescence glass dosimeters. *Radioisotopes* **2011**, *60*, 55–61. [CrossRef]
13. Li, B.; Sun, B.; Mu, H. High-efficiency generation of 355 nm radiation by a diode-end-pumped passively Q-switched Nd:YAG/Nd:YVO_4 laser. *Appl. Opt.* **2016**, *55*, 2474–2477. [CrossRef] [PubMed]
14. Yang, A. Experimental Demonstration of a Ring-Cavity Q-Switched Fiber Regenerative Amplifier. *Jpn. J. Appl. Phys.* **2006**, *45*, L673. [CrossRef]
15. Peng, Z.; Shi, Y.; Bu, X.; Hong, C.; Li, H.; Xu, Y.; Wang, P. 21 W, 105 μJ Regenerative Amplifier Based on Yb:YAG SCF and NALM Fiber Oscillator. *IEEE Photonics Technol. Lett.* **2020**, *32*, 333–336. [CrossRef]

16. Li, Y.; Tu, W.; Wang, B.; Zong, N.; Yang, F.; Zhang, F.; Zhang, S.; Wang, Z.; Peng, Q.; Cui, D.; et al. High-Power Narrow-Linewidth Q-Switched TEM00 Mode 355-nm Laser. *IEEE Photonics J.* **2018**, *10*, 1–9. [CrossRef]
17. Huang, Y.J.; Huang, Y.P.; Chiang, P.Y.; Liang, H.C.; Su, K.W.; Chen, Y.F. High-power passively Q-switched Nd:YVO4 UV laser at 355 nm. *Appl. Phys. B* **2011**, *106*, 893–898. [CrossRef]
18. Sakai, H.; Kan, H.; Taira, T. >1 MW peak power single-mode high-brightness passively Q-switched Nd3+: YAG microchip laser. *Opt. Express* **2008**, *16*, 19891–19899. [CrossRef]
19. Dörring, J.; Killi, A.; Morgner, U.; Lang, A.; Lederer, M.; Kopf, D. Period doubling and deterministic chaos in continuously pumped regenerative amplifiers. *Opt. Express* **2004**, *12*, 1759–1768. [CrossRef]
20. Zhu, P.; Li, D.; Liu, Q.; Chen, J.; Fu, S.; Shi, P.; Du, K.; Loosen, P. 39.1 μJ picosecond ultraviolet pulses at 355 nm with 1 MHz repeat rate. *Opt. Lett.* **2013**, *38*, 4716–4718. [CrossRef]
21. Borsutzky, A.; Brünger, R.; Huang, C.; Wallenstein, R. Harmonic and sum-frequency generation of pulsed laser radiation in BBO, LBO, and KD* P. *Appl. Phys. B* **1991**, *52*, 55–62. [CrossRef]
22. Yoshida, H.; Fujita, H.; Nakatsuka, M.; Yoshimura, M.; Sasaki, T.; Kamimura, T.; Yoshida, K. Dependences of Laser-Induced Bulk Damage Threshold and Crack Patterns in Several Nonlinear Crystals on Irradiation Direction. *Jpn. J. Appl. Phys.* **2006**, *45*, 766–769. [CrossRef]
23. Zhang, B.T.; Huang, H.T.; Yang, J.F.; He, J.L.; Zuo, C.H.; Xu, J.L.; Yang, X.Q.; Zhao, S. Generation of 7.8W at 355nm from an efficient and compact intracavity frequency-tripled Nd:YAG laser. *Opt. Commun.* **2010**, *283*, 2369–2372. [CrossRef]
24. Wang, X.; Wang, B.; Guo, S.; Cheng, W. Stable and compact nanosecond pulse seed source with diode laser driven by a high-speed circuit and a temperature control circuit. *Optik* **2019**, *179*, 1123–1128. [CrossRef]
25. Wang, G.; Liu, J.; Liu, S.; Liu, M.; Liu, S.; Li, L. Short pulse duration of a 355-nm laser by extracavity sum-frequency mixing with a LiB_3O_5 (LBO) crystal. *Laser Phys. Lett.* **2008**, *5*, 506–509. [CrossRef]

photonics

MDPI

Communication

Nonlinear-Optical-Loop-Mirror-Based Mode-Locked Fiber Laser Sensor for Low-Temperature Measurement

Xian-An Dou [1,2,3], Linchan Li [1,2,3], Chang Liang [4], Haiping Xu [1,2,3,*], Qing Ye [1,2,3], Hui Kong [1,2,3], Jintian Bian [1,2,3] and Lei Guo [4,*]

1 State Key Laboratory of Pulsed Power Laser Technology, National University of Defense Technology, Hefei 230037, China; xian.an_dou@nudt.edu.cn (X.-A.D.); lilinchan20@nudt.edu.cn (L.L.); yeqing18@nudt.edu.cn (Q.Y.); konghui17@nudt.edu.cn (H.K.); bianjintian17@nudt.edu.cn (J.B.)
2 Advanced Laser Technology Laboratory of Anhui Province, Hefei 230037, China
3 Nanhu Laser Laboratory, National University of Defense Technology, Changsha 410073, China
4 School of Mechanics and Optoelectronic Physics, Anhui University of Science and Technology, Huainan 232001, China; 2023201264@aust.edu.cn
* Correspondence: xuhaiping17@nudt.edu.cn (H.X.); leiguo@aust.edu.cn (L.G.)

Abstract: A temperature-sensing scheme is realized by a passively mode-locked Yb-doped fiber laser based on the nonlinear optical loop mirror (NOLM). The ambient temperature can be measured by detecting the pulse repetition frequency of the mode-locked fiber laser by an oscilloscope. When the ambient temperature increases from −40 °C to 6 °C, the pulse repetition frequency decreases linearly with a temperature sensitivity of 72.548 Hz/°C. The experimental results prove the feasibility of the mode-locked laser sensor operating in a low-temperature environment.

Keywords: fiber laser sensor; mode-locked laser; temperature sensor; Yb-doped

1. Introduction

Fiber-optic temperature sensors have important applications in industry, aerospace, biomedicine, and national security with the advantages of small size, high sensitivity, corrosion resistance, strong anti-electromagnetic interference, intrinsic safety, and long-distance detection [1–5]. According to the sensor type, traditional fiber-optic temperature sensors are mainly based on fiber Bragg gratings (FBGs) and fiber interferometric structures, such as Mach Zehnder interferometers (MZIs), Michelson interferometers (MIs), and Fabry–Perot interferometers (FPIs) [6–9]. Using the spectral demodulation technology, the temperature of the environment can be measured by detecting changes in transmitted or reflected spectral signals. However, such schemes rely on the expensive optical spectrum analyzer (OSA) with relatively low sensitivity. In recent years, fiber laser sensing schemes have emerged as a promising alternative for high-precision fiber temperature sensing with the development of Er^{3+} or Yb^{3+} ion-doped fiber lasers, with advantages such as compact size, narrow bandwidth, high signal-to-noise ratio (SNR), and good beam quality.

Compared with spectral demodulation technology, radio-frequency (RF) demodulation based on fiber lasers has the advantages of high speed, high resolution, and low cost [10]. The generation of multiple longitudinal modes is essential for these fiber laser sensors, typically achieved using a continuous-wave laser. A beat frequency signal (BFS) is generated between any two longitudinal modes, which is received by the photoelectric detector (PD) and converted into an electrical signal for display on the frequency spectrum analyzer (FSA). The BFS varies linearly with temperature, and temperature sensing is realized by measuring the frequency shift. Zuowei Yin et al. presented a novel multilongitudinal-mode (MLM) fiber ring laser sensor for temperature measurement with sensitivities of 10.24 kHz/°C @ 1581.7 MHz between 21 °C and 50 °C, 7.42 kHz/°C @ 1175.6 MHz between 22 °C and 240 °C, and 8.48 kHz/°C @ 1480.5 MHz between 100 °C and 1000 °C [11]. Long Huang et al. proposed an MLM fiber laser sensor with a sensitivity of −4.22 kHz/°C @ 1004.8 MHz and −16.78 kHz/°C @ 4041.2 MHz [12]. Xiujuan Yu et al. proposed a polarimetric MLM fiber laser with a sensitivity of −25.53 kHz/°C between 20 °C and 90 °C [13]. Xingxing Tong et al. proposed an MLM laser sensor system based on the neural network (NN) algorithm, demonstrating a single beat frequency sensitivity of 5.204 kHz/°C [14]. Generally, the higher the frequency of the beat signal, the higher the measurement accuracy. However, it is challenging for conventional lasers to generate a large number of longitudinal modes with a high SNR. Mode-locked lasers can generate plenty of longitudinal modes with equal frequency intervals, which have great potential for temperature, strain, and humidity monitoring in harsh environments [15–18]. Recently, Jian Luo et al. proposed a temperature-sensing scheme based on a passively mode-locked erbium-doped fiber (EDF) laser [19]. At a beat frequency of 10 GHz, the temperature sensitivity reaches up to −44 kHz/°C between 20 °C and 80 °C. Shaonian Ma et al. presented an optomechanically mode-locked EDF laser sensor in the temperature range of 20 °C to 65 °C, yielding a temperature sensitivity of −13.3 kHz/°C @ 1.917 GHz with a high SNR of 58 dB [20]. When the mode-locked fiber laser sensor is used outdoors, it is an important prerequisite for temperature sensing that mode-locked laser can work normally in high- and low-temperature environments. However, there are no reports on the temperature sensing characteristics of mode-locked fiber lasers in low-temperature environments.

In this paper, we propose a temperature-sensing scheme for low-temperature applications using a passively mode-locked fiber laser based on the nonlinear optical loop mirror (NOLM). A platinum resistance thermometer is employed in the heating device to monitor temperature variations. The temperature sensing demodulation system is a common digital oscilloscope. The variation in the pulse repetition frequency can reflect corresponding temperature changes. The experimental results demonstrate that the pulse repetition frequency decreases linearly with increasing temperature with a temperature sensitivity of 72.548 Hz/°C. Here, the pulse repetition frequency of the mode-locked laser corresponds to the fundamental frequency of the FSA. If the frequency spectrum analyzer with GHz bandwidth is used, the temperature sensitivity can be comparable to the current report.

2. Experimental Setup

Figure 1 shows the experimental setup of the mode-locked fiber laser used for temperature sensing. The pump source is a single-mode 980 nm laser diode (LD) with a maximum output power of 500 mW. A 980/1064 nm wavelength division multiplexer (WDM) connects the pump source to the resonator through the 980 nm port. The pump light is coupled into the upper unidirectional ring and propagates clockwise through the WDM. The gain medium is an 80 cm polarization-maintaining ytterbium-doped fiber (PM-YDF) (PM-YSF-HI-HP, Nufern, East Granby, CT, USA). After the gain fiber, an isolator (ISO) is used to ensure unidirectional transmission of the laser in the upper ring. In order to suppress the radiation around 1030 nm, a 1064 nm bandpass fiber filter with a 2 nm bandwidth is used. A 10/90 fiber coupler (OC2) connects the upper unidirectional ring cavity with the lower bidirectional Sagnac ring. To reduce the mode-locking threshold, a phase shifter (PS) is used to provide an initial phase bias of $\pi/2$. Using an asymmetrical fiber coupler and a polarization-maintaining single-mode fiber on the lower side, an NOLM-based passively mode-locked fiber laser is successfully constructed with a figure-8 cavity [21]. Another fiber coupler (OC1) has a coupling ratio of 30/70. The 30% port of OC1 is used for the laser output, while the 70% port is connected inside the cavity to maintain its normal operation.

The total cavity length is 27 m, and the length of the lower bidirectional Sagnac ring is 13 m. The waveform of the mode-locked pulse is monitored in real-time using a digital oscilloscope (InfiniiVision DSOX6002A, KEYSIGHT, Santa Rosa, CA, USA).

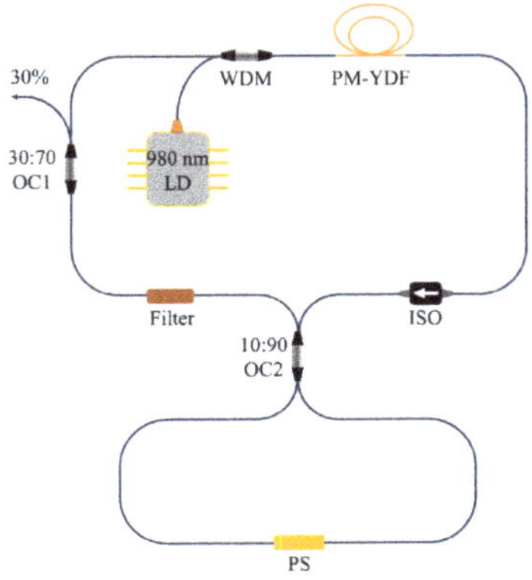

Figure 1. Experimental setup of the passively mode-locked figure-8 cavity laser.

3. Experimental Results

Firstly, the output characteristics of the mode-locked Yb-doped fiber laser is studied. The relationship between the average output power and the incident pump power is depicted in Figure 2. When the pump power is lower than 130 mW, the laser operates in the Q-switched mode-locking (QSML) regime. As the pump power continues to increase, continuous-wave mode-locking (CWML) is self-started with a center wavelength of ~1063 nm. The maximum average output power of 16 mW is obtained under an incident pump power of 400 mW. The pulse repetition frequency as a function of the incident pump power is plotted in Figure 3. Once the laser achieves stable mode-locking, the repetition frequency remains at 7.6562 MHz. Figure 4 illustrates the temporal pulse train of the mode-locked laser with a pump power of 220 mW. And the pulse repetition frequency is stable with a fluctuation of a few tenths of Hz. This indicates that the laser is suitable for further temperature sensing experiments.

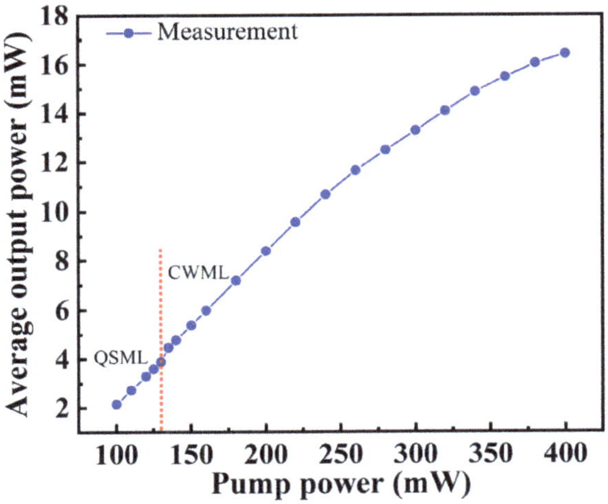

Figure 2. The average output power versus the incident pump power. The red dashed line represents the boundary between QSML and CWML.

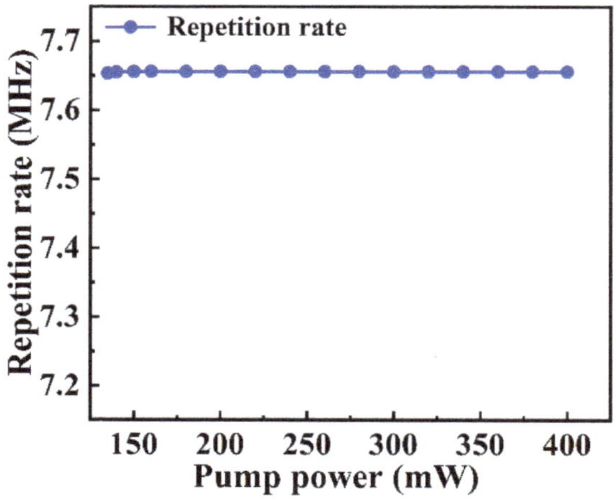

Figure 3. The relationship between the incident pump power and the repetition frequency.

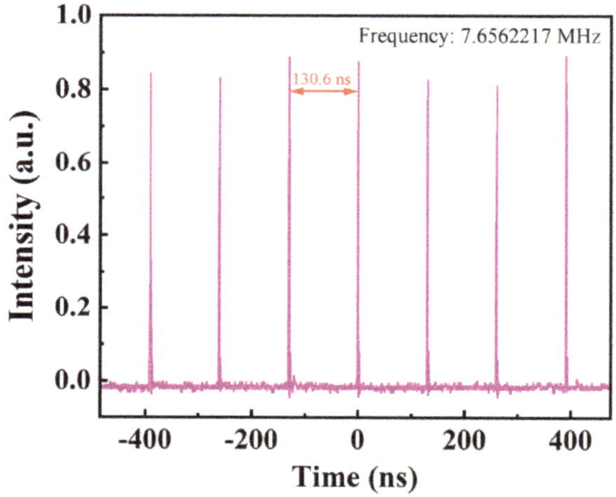

Figure 4. The mode-locked pulse sequence under an incident pump power of 220 mW. The red double headed arrow indicates a pulse spacing of 130.6 ns.

Subsequently, the mode-locked fiber laser is placed in a high- and low-temperature test chamber (ESS-KWGDS705 II, Chongqing Yinhe Experimental Equipment Co., Ltd., Chongqing, China). The temperature change range of the test chamber is from −70 °C to 150 °C, and the temperature change rate is ≤5 °C/min. The core device of the temperature sensor in our work is a passively mode-locked fiber laser based on the NOLM. By measuring the change in the pulse repetition frequency, the environment temperature can be accurately detected. In a mode-locked laser, the pulse repetition frequency for the ring fiber cavity is $f = c/nL$, where c, n, and L are the speed of light in vacuum, the refractive index, and the laser cavity length, respectively [22]. The temperature variation (ΔT) will affect the fiber length (L) and the refractive index (n) of the fiber. The derivative of the repetition frequency with respect to temperature is expressed as

$$\frac{df}{dT} = -c\frac{1}{nL}\left(\frac{1}{L}\frac{dL}{dT} + \frac{1}{n}\frac{dn}{dT}\right) = -f_r\left(\frac{1}{L}\frac{dL}{dT} + \frac{1}{n}\frac{dn}{dT}\right) \quad (1)$$

$$\Delta f = -f_r\left(\frac{1}{L}\frac{dL}{dT} + \frac{1}{n}\frac{dn}{dT}\right)\Delta T \tag{2}$$

where $\frac{1}{L}\frac{dL}{dT}$ is the thermal expansion coefficient and dn/dT is the thermo-optic coefficient of the optical fiber [23,24]. The laser repetition frequency at 0 °C is used as the reference frequency f_r. According to Equation (2), the pulse repetition frequency decreases linearly with the increase in temperature. The sensitivity of the temperature-sensing scheme is defined as the amount of change in the repetition frequency when the temperature changes by 1 °C. This sensitivity is defined as S (Hz/°C); thus, Equation (2) can be rewritten as

$$S = \frac{\Delta f}{\Delta T} = -f_r\left(\frac{1}{L}\frac{dL}{dT} + \frac{1}{n}\frac{dn}{dT}\right) \tag{3}$$

where the unit of f_r is MHz. It can be observed that the temperature sensitivity is related to the pulse repetition frequency, the thermal expansion coefficient, and the thermo-optic coefficient. Therefore, the sensitivity of the temperature sensor can be enhanced by increasing the repetition frequency of the mode-locked laser and using fibers with a high linear expansion coefficient.

In order to investigate the effect of environmental temperature variation on the pulse repetition frequency, the incident pump power is set to 220 mW. Considering the error between the actual temperature and the set temperature of the test chamber, a PT1000 platinum resistance thermometer with a resolution of 0.001 °C is used to measure the actual temperature value. As shown in Figure 5a,b, the relationship between the laser repetition frequency and the temperature as a function of time is recorded when the chamber temperature is set to −40 °C and 0 °C. It is clear that there is a highly negative correlation between laser pulse repetition frequency and temperature. At the same time, it also shows that the influence of ambient vibration on pulse frequency can be neglected in this experiment.

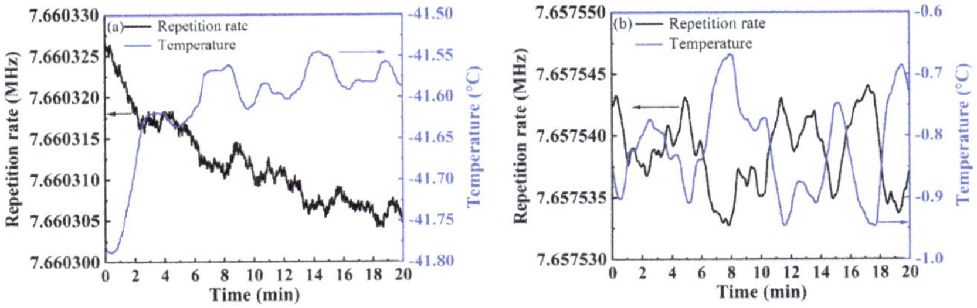

Figure 5. The temperature and repetition frequency as a function of time when the chamber temperature is set to (**a**) −40 °C and (**b**) 0 °C.

Figure 6a shows the relationship between the pulse repetition frequency and the temperature over time at low temperature. Since the temperature of the test chamber fluctuates greatly during the cooling process, we first lower the temperature to −40 °C in the experiment and then test the heating process. For the purpose of reducing the hysteresis effect of temperature measurement, the temperature rise rate of the test chamber should not rise too fast. As the temperature increases from −40 °C to 6 °C, the repetition frequency of the mode-locked laser gradually decreases. The repetition frequency as a function of temperature is plotted in Figure 6b with a linear decreasing trend. The temperature sensitivity is 72.548 Hz/°C by linear fitting. If the beat frequency demodulation system is used, a higher temperature sensitivity will be obtained.

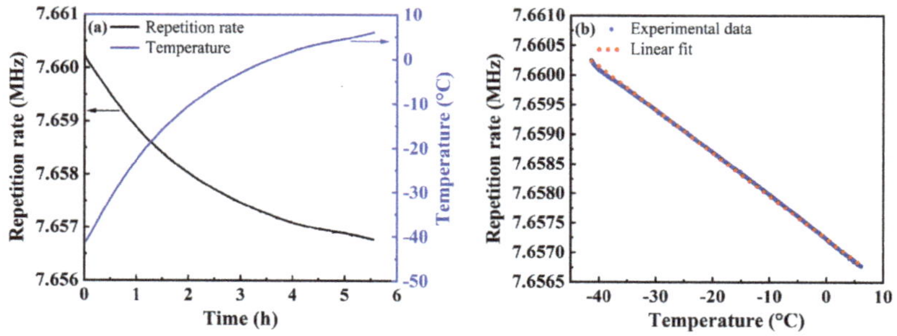

Figure 6. (**a**) The relationship between pulse repetition frequency and temperature over time and (**b**) the repetition frequency changes with temperature at low temperature.

To further verify the temperature resolution of the mode-locked fiber laser temperature sensor, the sensor is placed outdoors for experiments. As can be seen from Figure 7, the room temperature rises slowly in the morning over a period of 50 s. When the ambient temperature changes little, the temperature measurement value of the platinum resistance thermometer increases stepwise, while the repetition frequency of the mode-locked fiber laser decreases linearly and smoothly with the slow increase in temperature. Compared with the platinum resistance thermometer, it has a higher temperature measurement resolution and higher sensitivity.

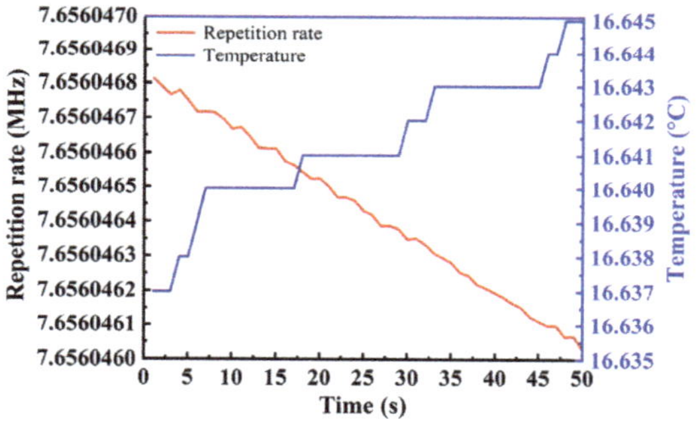

Figure 7. Comparison of temperature resolution between mode-locked fiber laser sensor and platinum resistance thermometer.

4. Conclusions

In summary, a novel fiber temperature-sensing scheme for low-temperature measurement is demonstrated based on a passively mode-locked fiber laser. The sensor is a 1063 nm mode-locked Yb-doped fiber laser with NOLMs. When the ambient temperature increases from −40 °C to 6 °C, the repetition frequency of the mode-locked laser gradually decreases with the temperature sensitivity of 72.548 Hz/°C. At the same time, this type of temperature sensor also shows high-temperature measurement resolution and high sensitivity. Compared to conventional fiber-optic temperature sensors, this approach enhances both the sensitivity and SNR of temperature-sensing technology. Furthermore, as an innovative temperature-sensing method, it possesses substantial potential for future advancements.

Author Contributions: Conceptualization, X.-A.D.; methodology, X.-A.D. and L.L.; validation, H.X.; form analysis, L.G. and J.B.; investigation, X.-A.D. and H.X.; resources, J.B.; data curation, L.G.; software, C.L.; visualization, H.X.; writing—original draft preparation, X.-A.D.; writing—review and editing, L.L.; supervision and funding acquisition, L.G., Q.Y. and H.K. All authors have read and agreed to the published version of the manuscript.

Funding: This work was supported by the National Natural Science Foundation of China (No. 62105003), Independent Research Special Project of National University of Defense Technology (Project No. ZZKY-YX-06-05), Advanced Laser Technology Laboratory Foundation of Anhui Province (AHL2022ZR03), Technology Domain Fund of Basic Strengthening Plan (2022-JCJQ-JJ-0237), and Independent Innovation Science Foundation of National University of Defense Technology (23-zzcx-jdz-44).

Institutional Review Board Statement: Not applicable.

Informed Consent Statement: Not applicable.

Data Availability Statement: Data underlying the results presented in this paper are not publicly available at this time but may be obtained from the author upon reasonable request.

Conflicts of Interest: The authors declare no conflicts of interest.

References

1. Shi, Y.T.; Li, X.Y.; Wang, P.; Zhang, Y.; Shi, B.H.; Li, Y.Y. A Review: Salinity and Temperature Measurement Based on Optical Fiber Sensors. *Sens. Actuators A Phys.* **2024**, *380*, 116040. [CrossRef]
2. Tangudu, R.; Sahu, P.K. Review on the Developments and Potential Applications of the Fiber Optic Distributed Temperature Sensing System. *IETE Tech. Rev.* **2022**, *39*, 553–567. [CrossRef]
3. Wang, L.; Wang, Y.J.; Song, S.M.; Li, F. Overview of Fibre Optic Sensing Technology in the Field of Physical Ocean Observation. *Front. Phys.* **2021**, *9*, 745487. [CrossRef]
4. Ma, S.N.; Xu, Y.P.; Pang, Y.X.; Zhao, X.; Li, Y.F.; Qin, Z.G.; Liu, Z.J.; Lu, P.; Bao, X.Y. Optical Fiber Sensors for High-Temperature Monitoring: A Review. *Sensors* **2022**, *22*, 5722. [CrossRef]
5. Paul, K.U.; Dhiman, D.; Muneesh, M.; Yang, Y.W.; Manojit, P. Real-time monitoring of temperature using a pulsed laser-diode-based photoacoustic system. *Opt. Lett.* **2020**, *45*, 718–721.
6. Khan, R.Y.M.; Ullah, R.; Faisal, M. Design and Development of Cost-Effective Fiber Bragg Grating Temperature Sensor Package. *Meas. Sci. Technol.* **2023**, *34*, 085122. [CrossRef]
7. González-Roque, A.A.; Toral-Acosta, D.; Martínez-Ríos, A.; Selvas-Aguilar, R.; Anzueto-Sánchez, G.; Rico-Méndez, M.A.; Guzmán-Ramos, V. Two-Mode Fiber Mach-Zehnder Interferometric Temperature Sensor in the 50 °C–650 °C Range. *Opt. Fiber Technol.* **2023**, *81*, 103568. [CrossRef]
8. Abbas, L.G.; Li, H. Temperature Sensing by Hybrid Interferometer Based on Vernier like Effect. *Opt. Fiber Technol.* **2021**, *64*, 102538. [CrossRef]
9. He, C.; Zhou, C.; Zhou, Q.; Xie, S.Y.; Xiao, M.Z.; Tian, J.J.; Yao, Y. Simultaneous Measurement of Strain and Temperature Using Fabry–Pérot Interferometry and Antiresonant Mechanism in a Hollow-Core Fiber. *Chin. Opt. Lett.* **2021**, *19*, 041201. [CrossRef]
10. Liu, Y.; Zhang, M.J.; Wang, P.; Li, L.; Wang, Y.C.; Bao, X.Y. Multiwavelength single-longitudinal-mode Brillouin–erbium fiber laser sensor for temperature measurements with ultrahigh resolution. *IEEE Photonics J.* **2015**, *7*, 6802809.
11. Yin, Z.W.; Gao, L.; Liu, S.C.; Zhang, L.; Wu, F.X.; Chen, L.; Chen, X.F. Fiber Ring Laser Sensor for Temperature Measurement. *J. Light. Technol.* **2010**, *28*, 3403–3408.
12. Huang, L.; Qian, L.; Chen, L.; Gao, L.; Chen, X.F. Multilongitudinal Mode Fiber Laser Sensor for Temperature Measurement. In Proceedings of the 2012 Asia Communications and Photonics Conference (ACP), Guangzhou, China, 7–10 November 2012; p. AF4B.27.
13. Yu, X.J.; Dong, X.; Chen, X.F.; Zhang, J.T.; Liu, S.C. Polarimetric Multilongitudinal Mode Fiber Laser for Simultaneous Measurement of Strain and Temperature. *J. Light. Technol.* **2016**, *34*, 4941–4947. [CrossRef]
14. Tong, X.X.; Shen, Y.X.; Mao, X.W.; Yu, C.; Guo, Y. Fiber-Optic Temperature Sensor Based on Beat Frequency and Neural Network Algorithm. *Opt. Fiber Technol.* **2022**, *68*, 102783. [CrossRef]
15. Pang, Y.F.; Zhang, X.Q.; Cao, Z.; Xu, L.J. Precise Gas Temperature Measurement Using a Single Dual-Wavelength Mode-Locked Fiber Laser. *IEEE Trans. Instrum. Meas.* **2024**, *73*, 7005211. [CrossRef]
16. Luo, J.; Cai, X.; Wang, H.R.; Fu, H.Y. Temperature Sensing Technique by Using a Microwave Photonics Filter Based on an Actively Mode-locked Fiber Laser. *Microw. Opt. Techn. Let.* **2021**, *63*, 2535–2540. [CrossRef]
17. Kbashi, H.J.; Sheil, B.B.; Perego, A.M. A Mode-Locked Fibre Laser Temperature Independent Strain Sensor Based on Intracavity Pulse Interference. *Opt. Laser Eng.* **2024**, *176*, 108040. [CrossRef]

18. Ma, S.N.; Pang, Y.X.; Ji, Q.; Zhao, X.; Qin, Z.G.; Liu, Z.J.; Xu, Y.P. Relative Humidity Sensing Based on Passively Mode-Locked Fiber Laser with Polyimide-Coated Fiber. *IEEE Sens. J.* **2024**, *24*, 6257–6263. [CrossRef]
19. Luo, J.; Wang, H.R.; Cai, X.; Luo, Z.Q.; Fu, H.Y. Temperature-Sensing Scheme Based on a Passively Mode-Locked Fiber Laser via Beat Frequency Demodulation. *Chin. Opt. Lett.* **2023**, *21*, 020603. [CrossRef]
20. Ma, S.N.; Pang, Y.X.; Ji, Q.; Zhao, X.; Qin, Z.G.; Liu, Z.J.; Lu, P.; Bao, X.Y.; Xu, Y.P. Optomechanically Mode-Locked Laser Sensor for High-Accuracy Temperature Measurement. *IEEE Sens. J.* **2024**, *24*, 311–317. [CrossRef]
21. Liang, X.; Wu, G.; Wang, Q.; Han, Y.; Gao, B.; Tian, X.J.; Guo, S.X.; Liu, L. Dynamics of Solitons in NOLM Based All-Normal Dispersion Yb-Doped Mode-Locked Fiber Laser. *Opt. Fiber Technol.* **2023**, *81*, 103507. [CrossRef]
22. Korobko, D.A.; Fotiadi, A.A.; Zolotovskii, I.O. Mode-Locking Evolution in Ring Fiber Lasers with Tunable Repetition Rate. *Opt. Express* **2017**, *25*, 21180. [CrossRef] [PubMed]
23. Rego, G. Temperature Dependence of the Thermo-Optic Coefficient of SiO_2 Glass. *Sensors* **2023**, *23*, 6023. [CrossRef] [PubMed]
24. Rego, G.M. Temperature Dependence of the Thermo-Optic Coefficient of GeO_2-Doped Silica Glass Fiber. *Sensors* **2024**, *24*, 4857. [CrossRef] [PubMed]

photonics

MDPI

Article

Exploring the Impact of Inlet Velocity Distribution on the Thermal Performance of a Laser Rod in a Diode Side-Pumped Amplifier

Shuzhen Nie [1,2,*], Jinglan Lin [1,2], Tianzhuo Zhao [2] and Xiaolong Liu [1,2,*]

1 Aerospace Information Research Institute, Chinese Academy of Sciences, Beijing 100094, China; linjinglan23@mails.ucas.ac.cn
2 School of Optoelectronics, University of Chinese Academy of Sciences, Beijing 100049, China; zhaotianzhuo@ucas.ac.cn
* Correspondence: niesz@aircas.ac.cn (S.N.); liuxiaolong@aircas.ac.cn (X.L.); Tel.: +86-010-8217-8661 (S.N.)

Abstract: Research on the thermal analysis of laser diode (LD) side-pumped amplifiers is a critical step in the design of high-power solid-state laser systems. Instead of adopting a standard solid modeling approach that only considers a laser rod, a fluid–structure interaction model is employed for analysis using the FLUENT 2021 R1 software. This model integrates the cooling structure, coolant, and laser rod, incorporating their relevant material parameters. By considering both uniform and non-uniform inlet velocity distributions as loading conditions, the study reveals remarkably different thermal simulation results. The correlation between thermal analysis outcomes and the total inlet flow rates is calculated, while temperature and stress distributions are obtained under a varying internal heat source. It was observed that the non-uniform inlet velocity distribution has little impact on the rod's maximum temperature but significantly influences the maximum equivalent stress. This finding underscores the necessity of accounting for non-uniform inlet distributions during the design of laser amplifiers to achieve more accurate thermal simulation results and optimize structural reliability.

Keywords: LD side-pumped laser amplifier; thermal analysis; inlet velocity distribution; the maximum temperature; the maximum equivalent stress

1. Introduction

Diode-pumped solid-state lasers (DPSSLs) are widely utilized in industrial processing, scientific research, communications, and other domains due to their high efficiency, compact size, lightweight design, and long service life [1–3]. DPSSLs predominantly employ the Master Oscillator Power Amplifier (MOPA) architecture, where the power amplifier significantly impacts the output beam quality [4,5]. In contemporary high-power DPSSL systems, multiple LD arrays are typically combined with side-pumped large-diameter laser rods for amplification [6,7]. This configuration is known for its simplicity and reliability, offering high pump energy, scalable optical aperture, and uniform pump power distribution along the rod's axial length, thus enabling high average output power or pulse energy. However, the primary obstacle to achieving superior output performance stems from the thermal effects induced by high-power pumping [8–10], which can degrade or even damage the gain medium. Consequently, thermal analysis assumes a critical role in the design of laser amplifiers, especially for large-diameter laser rods, where thermal management directly influences operational stability and component longevity.

https://doi.org/10.3390/photonics12060603

In the side-pumped configuration of multiple LD arrays, coolant is typically designed to flow around the laser rod's outer surface for efficient heat dissipation [11,12]. Conventional thermal analysis models for laser rods often employ the finite element method, imposing specific boundary conditions on the rod [13,14]. A common practice in such modeling is to apply a convective heat transfer coefficient as a boundary condition to the rod's surface to simulate coolant heat dissipation [15]. Notably, these models typically focus solely on the rod itself, excluding the coolant and cooling channel from the computational domain. However, the actual heat dissipation performance of the coolant is primarily governed by its physical properties and the geometric design of the cooling channel. In our prior study [16], we developed a fluid–structure interaction model using ANSYS FLUENT 2021 R1 software to analyze thermal distributions in a rod within a laser amplifier system. By integrating computational fluid dynamics and numerical heat transfer analysis [17–19], we systematically investigated the correlations between the cooling layer thickness, coolant velocity, inlet pressure, convective heat transfer coefficient, maximum temperature, and maximum equivalent stress in the rod. A key discovery is that the coolant velocity profile at the inlet significantly influences the stress distribution in large-diameter laser rods side-pumped by multiple LD arrays. This paper aims to provide a comprehensive explanation of these findings, offering insights into optimizing cooling system designs for high-power laser amplifiers.

2. Theory

The cooling mechanism of the laser rod by the coolant involves three coupled processes: coolant fluid flow and heat convection, interfacial heat transfer at the fluid–solid boundary, and heat conduction within the solid structure. The cooling fluid is typically modeled as an incompressible viscous fluid. The temperature distribution of the coolant is governed by the convection–diffusion equation, which degenerates into the heat conduction equation in the solid domain where the fluid velocity is zero. These theoretical foundations and assumptions are consistent with those detailed in reference [16]. Based on these theories, the coolant velocity and the temperature distribution of the laser rod can be simulated by using ANSYS FLUENT. By mapping the rod's temperature results to a static structural analysis module, the thermal stress distribution within the rod can be obtained. The thermal expansion phenomenon induced by a temperature rise in the laser rod can be described microscopically as the displacement of individual elements, which can be represented by U_x, U_y, and U_z. The relationship between the resulting strain and corresponding displacement can be expressed as [20]:

$$\begin{cases} \varepsilon_x = \frac{\partial U_x}{\partial x}, \varepsilon_y = \frac{\partial U_y}{\partial y}, \varepsilon_z = \frac{\partial U_z}{\partial z} \\ \gamma_{xy} = \frac{\partial U_x}{\partial y} + \frac{\partial U_y}{\partial x}, \gamma_{xz} = \frac{\partial U_x}{\partial z} + \frac{\partial U_z}{\partial x}, \gamma_{yz} = \frac{\partial U_y}{\partial z} + \frac{\partial U_z}{\partial y} \end{cases} \quad (1)$$

ε_i $(i = x, y, z)$ and γ_{ij} $(i, j = x, y, z\ \&\ i \neq j)$, respectively, represent the normal strain and shear strain in various directions. For sufficiently small deformations, Hooke's law is satisfied between stress and strain:

$$\begin{cases} \varepsilon_x = \frac{1}{E}\left[\sigma_x - v(\sigma_y + \sigma_z)\right] + \alpha\Delta T \\ \varepsilon_y = \frac{1}{E}\left[\sigma_y - v(\sigma_z + \sigma_x)\right] + \alpha\Delta T \\ \varepsilon_z = \frac{1}{E}\left[\sigma_z - v(\sigma_x + \sigma_y)\right] + \alpha\Delta T \\ \tau_{xy} = G\gamma_{xy},\ \tau_{zx} = G\gamma_{zx},\ \tau_{yz} = G\gamma_{yz} \end{cases} \quad (2)$$

where σ is the normal stress, τ is the shear stress, $G = E/[2(1+v)]$, E represents Young's modulus, α is the coefficient of thermal expansion, and v is Poisson's ratio.

In steady-state conditions, the forces in all directions must satisfy the mechanical equilibrium conditions:

$$\begin{cases} \frac{\partial \sigma_x}{\partial x} + \frac{\partial \tau_{xy}}{\partial y} + \frac{\partial \tau_{xz}}{\partial z} + F_x = 0 \\ \frac{\partial \tau_{xy}}{\partial x} + \frac{\partial \sigma_y}{\partial y} + \frac{\partial \tau_{yz}}{\partial z} + F_y = 0 \\ \frac{\partial \tau_{xz}}{\partial x} + \frac{\partial \tau_{yz}}{\partial y} + \frac{\partial \sigma_z}{\partial z} + F_z = 0 \end{cases} \quad (3)$$

where F_i $(i = x, y, z)$ is the external force applied to a specific element within the laser rod. In conclusion, by integrating the above equations with appropriate boundary conditions, the stress distribution within the rod can be derived.

3. Configuration and Modeling

The side-pumped amplifier's pumping scheme is illustrated in Figure 1a. The rod's diameter is 18 mm, surrounded by a 2 mm thick cooling layer, and the glass tube has a wall thickness of 2 mm. The effective pumping length is 160 mm, while the total rod length is 200 mm. The structure of multiple annular LD arrays with a wavelength of 802 ± 1 nm is optimized for high-power pumping, and we have studied the realization of uniform fluorescence distribution within the laser rod [21]. In this paper, the detected fluorescence distribution is shown in Figure 1b.

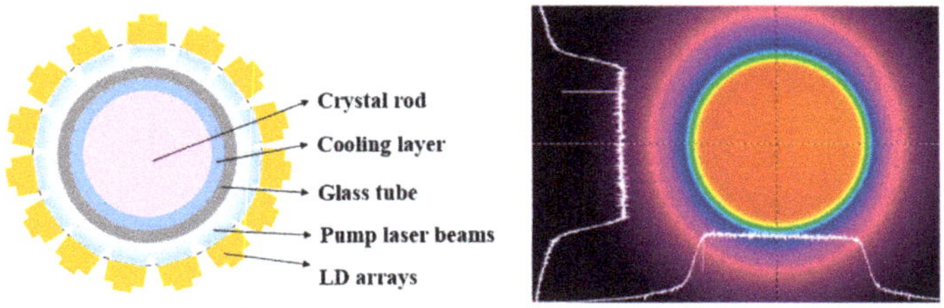

(a) (b)

Figure 1. (**a**) The pumping scheme of a side-pumped laser amplifier; (**b**) the detected fluorescence distribution.

A fluid–structure interaction model consisting of a glass tube, cooling channel, coolant, and laser rod was established in ANSYS FLUENT, as illustrated in Figure 2. The cooling channel is filled with coolant, which enters through one end of the glass tube and exits through the opposite end. Both ends are connected to external structure modules to facilitate coolant circulation. As a result, only the part of the rod enclosed within the glass tube is cooled by the coolant, while the other part is exposed to air to accommodate the laser rod's fixed support structure.

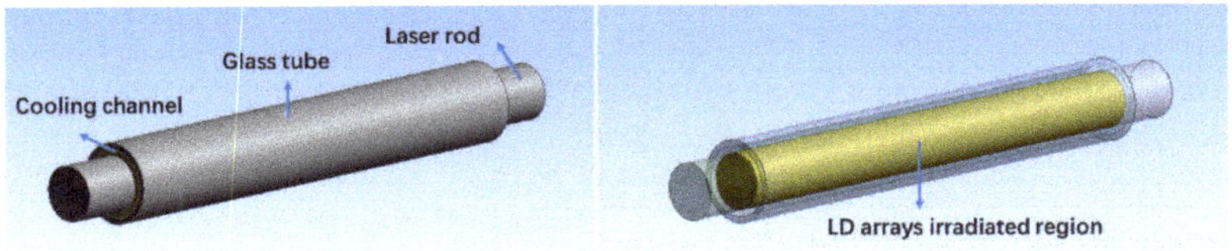

(a) (b)

Figure 2. The fluid–structure interaction model. (**a**) Three-dimensional model; (**b**) LD-arrays-irradiated region in the model.

In the LD-arrays-irradiated region of the laser rod, the absorbed pump laser power is treated as heat power. Generally, the conversion coefficient η is defined as the ratio of pump energy converted to heat power, with a value of 45% used for the laser rod. The LD arrays adopt a pulse pumping working mode and the pulse width, repetition frequency, and pump power are represented by τ, f, and P_{pump}, respectively. We assume that the heat power within the rod is uniform for simplicity, and an average pump power is used in order to reduce the simulation time. Therefore, the heat power can be calculated by $P_{pump} \cdot \eta \cdot \tau \cdot f$. We divide it by the volume of the irradiated region as the value of internal heat source in the model. The simulation results in Section 4.1. are calculated when the internal heat source is estimated to be 1.29×10^6 W/m^3, corresponding to pump parameters of a 500 μs pulse width, 1 Hz repetition frequency, and 234 kW pump power. The input and output conditions of the simulation model are the fluid velocity inlet and the pressure outlet. Rod surfaces in contact with the coolant adhere to the fluid–structure interaction boundary conditions at the fluid–solid interface, while surfaces without coolant adopt the boundary conditions of natural convection heat exchange with air. The model's initial temperature is typically set to room temperature 293.15 K. Gravity is also taken into account. The laser rod is a type of Nd glass (N31, Shanghai Institute of Optics and Fine Mechanics, Chinese Academy of Sciences, China), and the coolant is a kind of electronic fluorinated liquid (FC-770, 3M™ Fluorinert™, USA). Key model parameters are listed in Table 1 [22,23].

Table 1. Parameters used for the model.

Parameter (Materials)	Value (Units)
Density (laser rod)	2870 kg/m^3
Thermal conductivity (laser rod)	0.63 W/m/K
Specific heat (laser rod)	680 J/kg/K
Coefficient of thermal expansion (laser rod)	1.27×10^{-5} /K
Young's modulus (laser rod)	5.85×10^{10} Pa
Poisson's ratio (laser rod)	0.232
Density (coolant)	1793 kg/m^3
Thermal conductivity (coolant)	0.063 W/m/K
Specific heat (coolant)	1038 J/kg/K
Viscosity (coolant)	1.359×10^{-3} kg/m/s
Density (glass tube)	2200 kg/m^3
Thermal conductivity (glass tube)	1.4 W/m/K
Specific heat (glass tube)	670 J/kg/K

According to the Reynolds number $N_{Re} = (D_2 - D_1)\text{v}/\mu$ of the coolant in the cooling channel, where v and μ denote the velocity and viscosity of the coolant, the flow of coolant inside the cooling channel can be considered turbulent. Using the FLUENT software, the solving process can be carried out by selecting appropriate simulation conditions [24]. The turbulent flow inside the cooling channel was simulated employing the realizable k-ε model and standard wall functions. The velocity and pressure fields are calculated using the pressure-based coupling algorithm. The second-order upwind scheme is used to improve the calculation accuracy. Through transient fluid–structure coupling calculations, the simulation results of the thermal characteristics and fluid velocity distribution of the system after achieving thermal equilibrium can be obtained.

4. Simulation Results

4.1. *Unform Inlet Velocity Distribution*

Figure 3a displays the inlet area of the cooling channel. The simulation results indicate that at a coolant velocity of 0.8 m/s, the velocity distribution on the inlet surface is unform, as shown in Figure 3b, with the maximum temperature of 327 K occurring at the center of the LD-arrays-irradiated region (Figure 4a). The maximum temperature on the outer surface of the rod is 311 K, as shown in Figure 4b. The coolant velocity distribution in the channel is shown in Figure 4c, which presents a decreasing change from the inlet to the outlet direction. The temperature distribution on the outer surface of the rod is shown in Figure 4d, which is an increasing change from the inlet to the outlet direction. The temperature trend is directly correlated with variations in coolant velocity: regions with higher coolant velocity exhibit more efficient heat dissipation and lower temperatures. The maximum equivalent stress is 10.6 Ma, as illustrated in Figure 4e. The equivalent stress distribution along the z-axis is shown in Figure 4f, and it indicates that the regions where the rod is supported are subjected to greater stress. While the above simulation assumes a uniform inlet velocity distribution, practical cooling structures often impose flow restrictions, leading to non-uniform inlet velocity distributions. Investigating such non-uniform conditions is therefore critical for advancing laser amplifier design and improving thermal management.

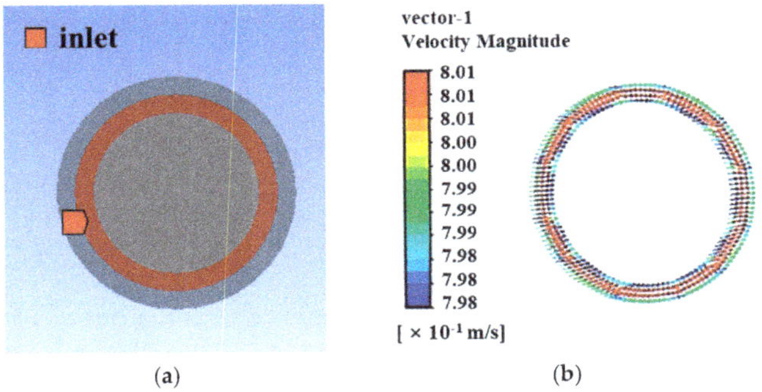

Figure 3. (**a**) The inlet area; (**b**) the velocity result on the inlet surface.

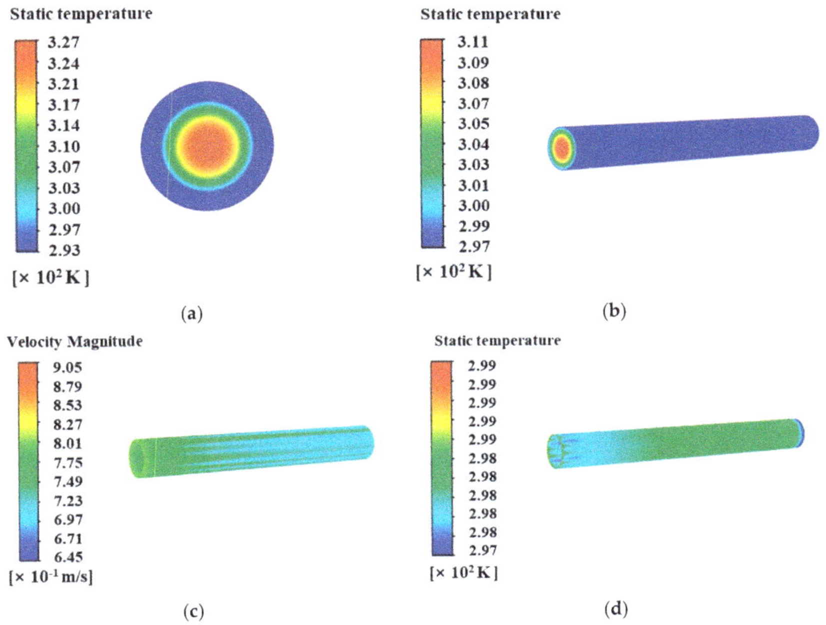

Figure 4. *Cont.*

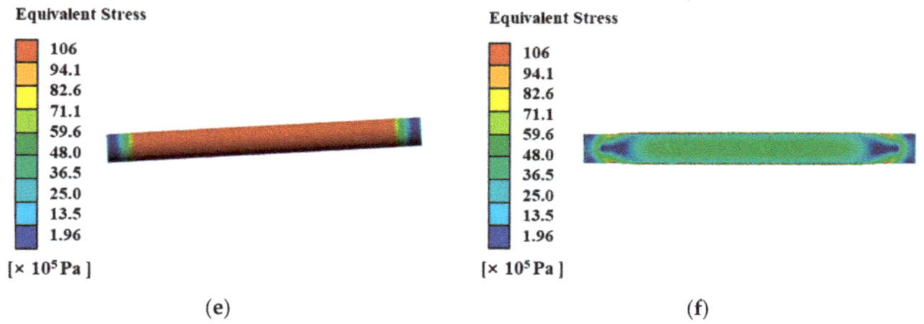

Figure 4. The simulation results with unform inlet velocity distribution: (**a**) the temperature profile in the center cross-section of the laser rod; (**b**) the temperature distribution on the outer surface of the laser rod; (**c**) the coolant velocity distribution in the cooling channel; (**d**) the temperature distribution on the outer surface of LD-arrays-irradiated region of the laser rod; (**e**) the equivalent stress result of the laser rod; (**f**) the equivalent stress distribution along the z-axis within the laser rod.

4.2. Non-Uniform Inlet Velocity Distribution

To characterize the effect of non-uniform inlet velocity distribution, the inlet surface is divided into six parts according to the cooling structure design, as shown in Figure 5a. The coolant flows from the cooling structure into the cooling channel through these six input regions. Different amplifiers have their own inlet structures, so the inlet area can be treated in different ways, for example, via continuous flow rate modulation. Here, the velocity decreases gradually from inlet1 to inlet6, and the total inlet flow rate can be calculated by the following:

$$u_{ft} = \sum_{i=1}^{6} (s_i \cdot v_i) \tag{4}$$

where s_i and v_i are the area and inlet velocity of each part. While ensuring the same total flow rate with the partitioning, the uniform equivalent velocity at the inlet surface (without partitioning) is defined as:

$$v_{eq} = u_{ft} / \sum_{i=1}^{6} s_i \tag{5}$$

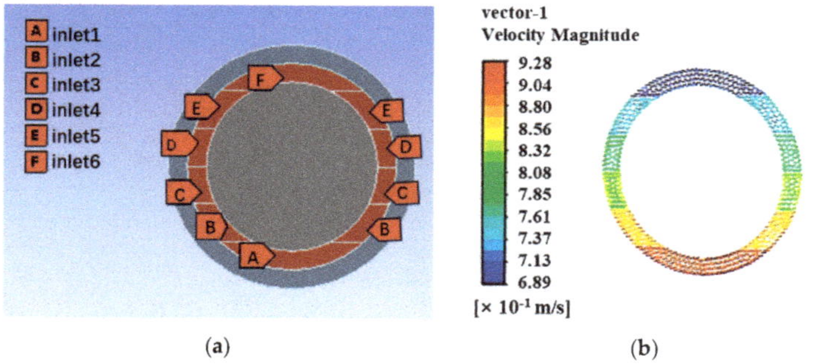

Figure 5. (**a**) The six parts at the inlet area; (**b**) the velocity result on the inlet surface.

The variation in the maximum or minimum velocity on the inlet surface relative to the equivalent velocity is approximately no more than 20%, which means $(v_{inlet\_\max/\min} - v_{eq})/v_{eq} \leq 20\%$, as proven by the previous simulation. When $v_{eq} = 0.8\ m/s$, with the same total flow rate of 6 L/min and $(v_{inlet\_\max/\min} - v_{eq})/v_{eq} = 12.5\%$, the velocity in each part from inlet1 to inlet6 is set as 0.9, 0.86, 0.82, 0.78, 0.74, and 0.7 m/s. The other solving conditions are the same as those without the partitioning. We hope to use this velocity variation on the inlet surface as a simplified means to investigate some different thermal characteristics. The simulated velocity result on the inlet surface is shown in Figure 5b, which matches the predefined profile. The temperature profile in the center cross-section remains nearly unchanged, as shown in Figure 6a. Other results exhibit some changes, with a non-uniform temperature distribution on the outer surface of the laser rod and an inconsistent coolant velocity distribution in the cooling channel. As shown in Figure 6b, although the maximum temperature value on the outer surface of the laser rod has remained basically unchanged, the temperature distribution on the outer surface of LD arrays irradiated region is not uniform, with some fluctuations that can be more clearly observed in Figure 6d. The coolant velocity distribution in the cooling channel is shown in Figure 6c. Compared to the results in Figure 4c, the velocity values are generally lower. The lower velocity is more likely to cause uneven heat dissipation, which may be the reason for the temperature fluctuations and significant change in terms of the maximum equivalent stress (Figure 6e). In Figure 6f, it also can be seen that the equivalent stress values near the outer surface are greater than at other parts. The maximum equivalent stress has increased to 20.1 MPa, almost doubling the previous value of 10.6 MPa. These results demonstrate that non-uniform inlet velocity distributions lead to a substantial increase in the maximum equivalent stress, while the maximum temperature of the rod remains the same.

To enhance heat dissipation, higher coolant inlet velocities are typically employed. Thermal simulation results are presented in Figure 7 with varying inlet total flow rates. In the case of non-uniform inlet velocity distributions, the velocity decreases equally from inlet1 to inlet6, with inlet1 (v_{inlet1}) having the highest velocity, maintaining the same ratio $(v_{inlet1} - v_{eq})/v_{eq}$ at different flow rates. At a constant total flow rate, there are differences in the trends in the maximum temperature and maximum equivalent stress between the uniform and non-uniform inlet velocities. At an inlet total flow rate of 6 L/min, the maximum temperature with a non-uniform inlet velocity is slightly higher than the uniform case. Conversely, at inlet total flow rates exceeding 10 L/min, the maximum temperature with a non-uniform inlet velocity is consistently slightly lower than the results with a uniform inlet velocity. This indicates that a non-uniform inlet velocity can reduce the maximum temperature rise of the laser rod to some extent at higher flow rates. The maximum equivalent stress results show that as the flow rate increases, the maximum equivalent stress remains nearly consistent under a uniform inlet velocity, whereas it decreases slightly under a non-uniform inlet velocity. Notably, substantial differences exist in the maximum equivalent stress between the two velocity conditions. In most cases, the assumption of a uniform inlet flow velocity can result in a significant discrepancy in the predicted maximum equivalent stress, compared to the actual non-uniform inlet velocity profile. This could be a contributing factor to the tendency of laser rods to fracture in practical applications.

Simulations of the maximum temperature and maximum equivalent stress with an increasing internal heat source are shown in Figure 8. At an inlet total flow rate of 6 L/min, the maximum temperature with a non-uniform inlet velocity remains comparable to the uniform case as the internal heat source increases, indicating that at lower flow rates, a non-uniform inlet velocity has a negligible effect on the maximum temperature. Conversely, the impact of a non-uniform inlet velocity on the maximum equivalent stress is notably higher than that of the uniform conditions. The results reveal that as the internal heat source increases, the maximum equivalent stresses under a uniform and non-uniform inlet velocity show a similar increasing trend but with a widening gap between them. When the internal heat source is 2.1×10^6 W/m^3, the maximum equivalent stress with a non-uniform inlet velocity exceeds the laser rod's fracture limit (51 MPa), predicting that the laser rod would break under these thermal loads. In contrast, the maximum equivalent stress with a uniform inlet velocity only reaches approximately 25 MPa.

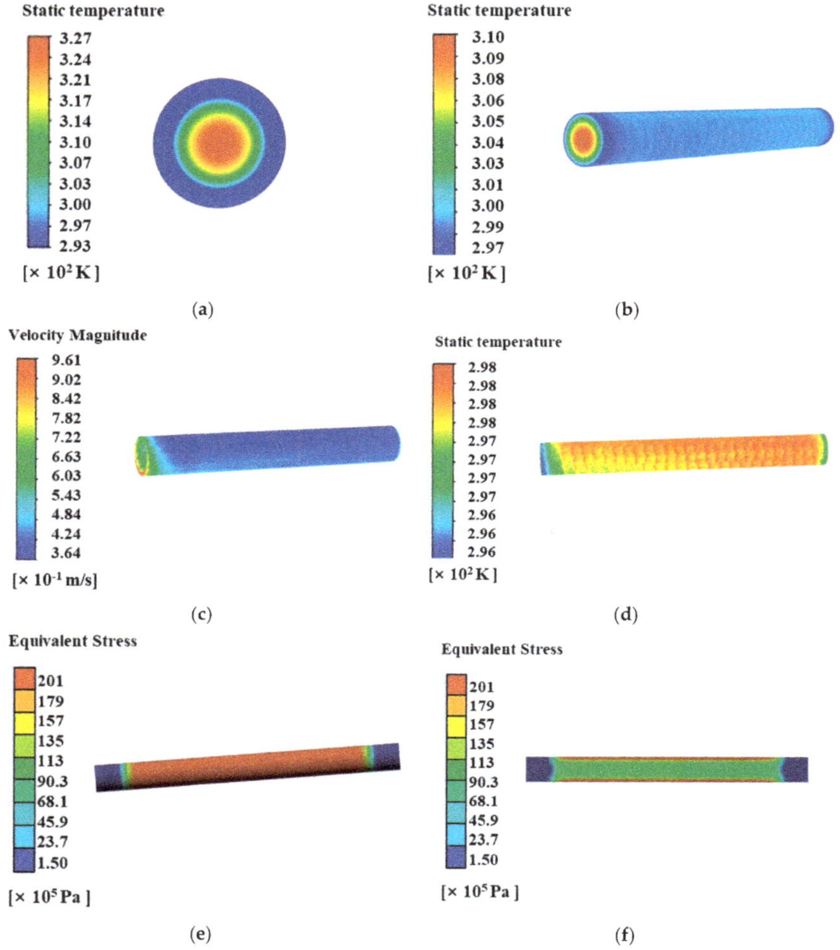

Figure 6. The simulation results with non-unform inlet velocity distribution: (**a**) the temperature profile in the center cross-section of the laser rod; (**b**) the temperature distribution on the outer surface of the laser rod; (**c**) the coolant velocity distribution in the cooling channel; (**d**) the temperature distribution on the outer surface of LD-arrays-irradiated region of the laser rod; (**e**) the equivalent stress result of the laser rod; (**f**) the equivalent stress distribution along the z-axis within the laser rod.

In high-power laser amplifiers, inlet flow rates exceeding 6 L/min are commonly used to optimize heat dissipation. When the inlet flow rate is 15 L/min, Figure 9 illustrates the simulated results under varying internal heat source conditions. The maximum temperatures increase with the internal heat source for both inlet velocity distributions, though minor differences exist between the two cases. This implies that non-uniform inlet velocity distributions at higher flow rates have a marginal effect on the maximum temperature. With an increase in the internal heat source, the maximum equivalent stress also exhibits an upward trend, but the growth rate varies. The non-uniform inlet velocity distributions have a more severe impact on the maximum equivalent stress. When the internal heat source is 2.2×10^6 W/m^3, the laser rod will fracture due to exceeding its fracture limit. In contrast, under a uniform inlet velocity, the maximum equivalent stress remains at only about 27 MPa below the failure threshold.

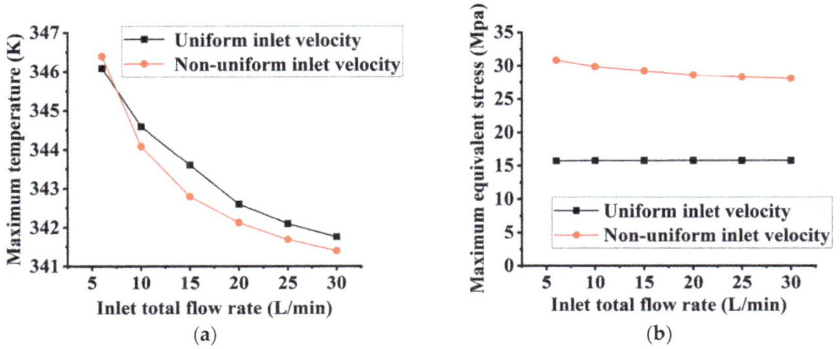

Figure 7. The results under different inlet total flow rate: (**a**) maximum temperature results; (**b**) maximum equivalent stress results.

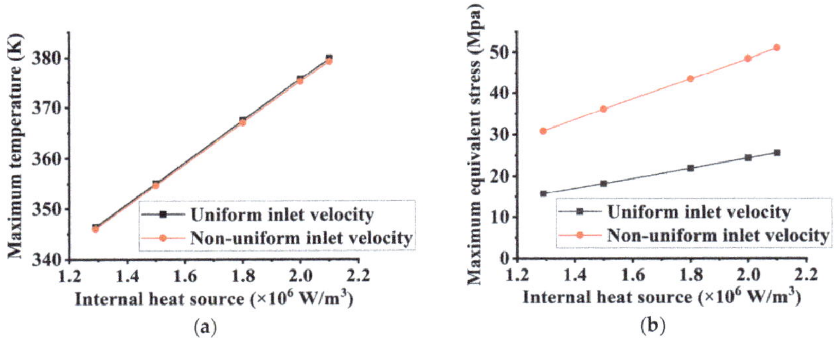

Figure 8. The results with different internal heat sources when the flow rate is 6 L/min: (**a**) maximum temperature results; (**b**) maximum equivalent stress results.

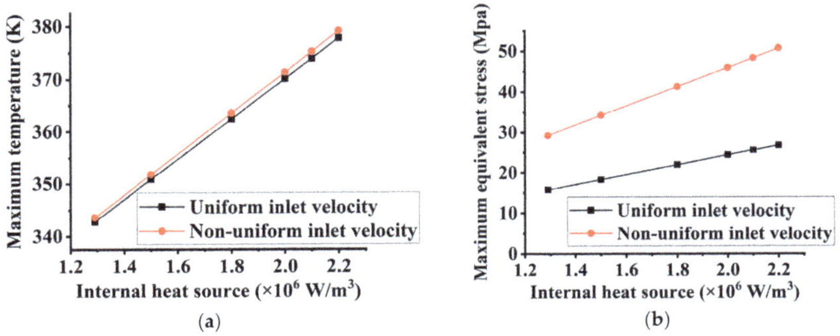

Figure 9. The results with different internal heat sources when the flow rate is 15 L/min: (**a**) maximum temperature results; (**b**) maximum equivalent stress results.

Accompanied by the enhancement of internal heat sources, the simulation results in Figure 10 compare and analyze the effects of different flow rates on heat dissipation with a non-uniform inlet velocity. Increasing the coolant flow rate from 6 L/min to 15 L/min enhances the heat dissipation efficiency, leading to a lower maximum temperature and equivalent stress. The rate of changes in both temperature and stress, with an increasing internal heat source, are similar between the two flow rates. When the internal heat source is 2.1 × 10^6 W/m^3, the maximum equivalent stress at 6 L/min exceeds 51 MPa, causing the laser rod to fracture under such loading conditions. However, with a flow rate of 15 L/min, the maximum equivalent stress remains below 50 MPa, and no fracture occurs. As the internal heat rate reaches 2.2 × 10^6 W/m^3, both flow rates result in maximum equivalent stresses exceeding 51 MPa, leading to the occurrence of fracture.

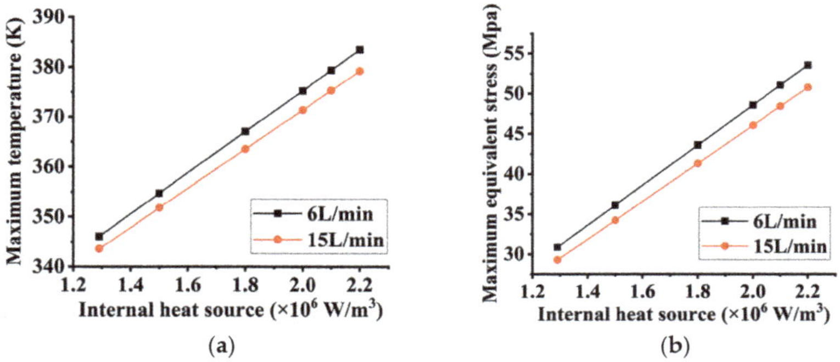

Figure 10. The results of different flow rates with non-uniform inlet velocity: (**a**) maximum temperature results; (**b**) maximum equivalent stress results.

5. Conclusions

In conclusion, through detailed presentation of the temperature and stress simulation results under both uniform and non-uniform inlet velocity distribution conditions, this study reveals that the inlet velocity distribution significantly influences the maximum equivalent stress in laser rods. These findings highlight the necessity of utilizing a fluid–structure interaction model that includes a glass tube, cooling channel, coolant, and laser rod for a more accurate simulation of the thermal performance of side-pumped laser amplifiers, rather than relying solely on standalone laser rod thermal models. By incorporating a non-uniform inlet velocity into the solving conditions, more accurate thermal performance results can be achieved. At higher flow rates, the non-uniform inlet velocity distribution has a negligible effect on the maximum temperature reduction. However, it can induce approximately twice the maximum equivalent stress compared to the uniform case. With the increase in the internal heat source, there is only a slight difference in the maximum temperature results between the two types of inlet velocity distributions, but the maximum equivalent stress results still exhibit significant differences. Conventional simulations often assume uniform inlet velocity distributions for simplicity, but actually, the distribution is non-uniform due to the complex cooling channel structure. When using a uniform inlet velocity as the input condition, the stress results obtained are generally lower than that with a non-uniform inlet velocity distribution, despite similar temperature outcomes. Therefore, it is possible that even if the simulated stress results under uniform conditions lie below the fracture limit, in practical applications under the same thermal loading, the laser rod may still break. This could be attributed to the non-negligible impact of the non-uniform inlet velocity distribution. Hence, in the designing process of laser amplifiers, the actual inlet velocity distribution should be considered, and a fluid–structure interaction model is preferred for more accurate simulation results.

Author Contributions: S.N. performed the modeling and simulation, conducted the analysis of the results, and wrote the draft; J.L. carried out data results verification and code writing; T.Z. provided the software and revised the draft; X.L. gave suggestions on project management and provided helpful discussions on the results. All authors have read and agreed to the published version of the manuscript.

Funding: This research was funded by the second open project of the Palace Museum (No. 202405008, which received public funding from Longfor-Palace Museum Cultural Fund and The Forbidden City Cultural Heritage Conservation Foundation) and Youth Innovation Promotion Association CAS.

Institutional Review Board Statement: Not applicable.

Informed Consent Statement: Not applicable.

Data Availability Statement: Data is contained within the article.

Conflicts of Interest: The authors declare no conflicts of interest.

References

1. Divoký, M.; Pilař, J.; Hanuš, M.; Navrátil, P.; Denk, O.; Severová, P.; Mason, P.; Butcher, T.; Banerjee, S.; De Vido, M.; et al. 150 J DPSSL operating at 1.5 kW level. *Opt. Lett.* **2021**, *46*, 5771–5773. [CrossRef] [PubMed]
2. Xu, J.; Kong, M.; Lin, A.; Song, Y.; Han, J.; Xu, Z.; Wu, B.; Gao, S.; Deng, N. Directly modulated green-light diode-pumped solid-state laser for underwater wireless optical communication. *Opt. Lett.* **2017**, *42*, 1664–1667. [CrossRef] [PubMed]
3. Koechner, W. *Solid-State Laser Engineering*; Springer: New York, NY, USA, 2006.
4. Nomura, J.; Hirosawa, K.; Yanagisawa, T.; Ohata, N.; Imamura, S.; Sakaizawa, D.; Tomii, N. Single-frequency 45-mJ pulses from a MOPA system using an Er,Yb:glass planar waveguide amplifier and a large mode area Er-doped fiber amplifier. *Opt. Lett.* **2023**, *48*, 1758–1761. [CrossRef] [PubMed]
5. Kuznetsov, I.; Mukhin, I.; Palashov, O.; Ueda, K.-I. Thin-rod Yb:YAG amplifiers for high average and peak power lasers. *Opt. Lett.* **2018**, *43*, 3941–3944. [CrossRef] [PubMed]
6. Zhao, T.; Fan, Z.; Xiao, H.; Huang, K.; Bai, Z.; Ge, W.; Zhang, H. Realizing Gaussian to flat-top beam shaping in traveling-wave amplification. *Opt. Express* **2017**, *25*, 33226–33235. [CrossRef]
7. Nie, S.; Zhao, T.; Fan, Z.; Hong, X. Thermal effect of large-diameter laser rod in multi-direction side-pump high power pulsed Nd:YAG laser amplifier. In Proceedings of the International Symposium on Optoelectronic Technology and Application 2016, Beijing, China, 9–11 May 2016; p. 101520H.
8. Ebadian, H.; Maleki, A. Efficient thermal induced birefringence compensation in a QCW diode pumped Nd:YAG laser. *Optik* **2018**, *165*, 362–369. [CrossRef]
9. Huang, W.; Guo, J.; Lu, X.; Wang, J.; Fan, W.; Li, X.; Zhu, J. Diode-end-pumped all-liquid-cooled square rod Nd: Glass amplifier with sub-joule energy extraction and ultralow thermal induced wavefront. *Opt. Laser Technol.* **2022**, *152*, 108084. [CrossRef]
10. Guoying, F.; Qunfei, O.; Jianguo, C.; Mingzhong, L. Simulation of the thermal effects in diode-pumped rod laser. In Proceedings of the Optical Science and Technology, SPIE's 48th Annual Meeting, San Diego, CA, USA, 3–8 August 2003; pp. 43–48.
11. Ye, X.; Xu, X.; Ren, H.; Zhang, B.; Lu, Y.; Chen, X.; Zhang, L.; Wei, X.; Wan, M.; He, T. Enhanced high-slope-efficiency and high-power LD side-pumped Er:YSGG laser. *Appl. Opt.* **2019**, *58*, 9949–9954. [CrossRef] [PubMed]
12. Zhang, Y.; Gu, H.; Cui, L.; Jiang, J.; Hao, H. Research on the LD side pumping field of solid-state lasers. In Proceedings of the SPIE Security + Defence, Strasbourg, France, 9–12 September 2019; Volume 11162.
13. Wang, W.; Zhao, Q.; Gao, W.; Hu, Z.; Zhao, Q.; Yang, S. Thermo-Optic Numerical Research on Segmented Circular LD Arrays Side-Pumping a Nd:YAG Laser Rod. *Appl. Sci.* **2020**, *10*, 7316. [CrossRef]
14. Weber, R.; Neuenschwander, B.; Weber, H.P. Thermal effects in solid-state laser materials. *Opt. Mater.* **1999**, *11*, 245–254. [CrossRef]
15. Nie, S.; Zhao, T.; Xiao, H.; Zhao, S. Influence of model characteristic parameters on thermal effect analysis of solid-state laser. In Proceedings of the 24th National Laser Conference & Fifteenth National Conference on Laser Technology and Optoelectronics, Shanghai, China, 17–20 October 2020.
16. Nie, S.; Zhao, T.; Liu, X.; Qu, P.; Yang, Y.; Wang, Y. The Effect of Cooling Layer Thickness and Coolant Velocity on Crystal Thermodynamic Properties in a Laser Amplifier. *Micromachines* **2023**, *14*, 299. [CrossRef] [PubMed]
17. Hsu, S.T. *Engineering Heat Transfer*; D. Van Nostrand Company: Princeton, NJ, USA, 1963.
18. Tao, W. *Numerical Heat Transfer*; Xi'an Jiaotong University Press: Xi'an, China, 2001.
19. Yang, S.T.W. *Heat Transfer*; Higher Education Press: Beijing, China, 2006.
20. Ding, J.; Li, Y.; Chen, H.; Cai, Y.; Bai, Z.; Qi, Y.; Yan, B.; Wang, Y.; Lu, Z. Thermal modeling of an external cavity diamond Raman laser. *Opt. Laser Technol.* **2022**, *156*, 108578. [CrossRef]
21. Xiao, H.; Zhao, T.; Fan, Z.; Huang, K.; Yu, J. Pump Design of Large-Diameter Axial Non-Uniform Nd:YAG Crystal. *Chin. J. Lasers* **2015**, *42*, 0602002. [CrossRef]
22. Wang, X.; Hu, L.; Chen, S.; Tang, J.; Meng, T.; Chen, W.; Weng, Z.; Chen, H.; Hu, J.; Chen, Y.; et al. Review of Development of Nd-Doped Laser Glass in China. *Chin. J. Lasers* **2024**, *51*, 1101008.

23. Li, J. Study on thermal design and thermal analysis of liquid cooling plate with FC-770. *Comput. Meas. Control* **2023**, *31*, 160–165.
24. Wu, G.; Zhao, X.; Shi, D.; Wu, X. Analysis of Fluid–Structure Coupling Vibration Mechanism for Subsea Tree Pipeline Combined with Fluent and Ansys Workbench. *Water* **2021**, *13*, 955. [CrossRef]

photonics MDPI

Article

Visible Triple-Wavelength Switchable Emission Generated in Passively Q-Switched Nd:YVO_4 Self-Raman Laser

Songtao Li [1,2], Shengxi Zheng [1,2], Bowen Zheng [1,2], Yong Wei [3,*], Yongchang Zhang [1,2], Yanmin Duan [1] and Haiyong Zhu [1,2,*]

1 Zhejiang International Sci-Tech Cooperation Base for Photoelectric Functional Devices and Digital Detections, Wenzhou University, Wenzhou 325035, China; 24451250009@stu.wzu.edu.cn (S.L.); 21211741117@stu.wzu.edu.cn (S.Z.); 23211710215@stu.wzu.edu.cn (B.Z.); yczhang@wzu.edu.cn (Y.Z.); ymduan@wzu.edu.cn (Y.D.)
2 Zhejiang Key Laboratory of Smart Low-Voltage Apparatus and New Energy Application, College of Electrical and Electronic Engineering, Wenzhou University, Wenzhou 325035, China
3 Key Laboratory of Hunan Province on Information Photonics and Freespace Optical Communications, College of Physics and Electronics, Hunan Institute of Science and Technology, Yueyang 414006, China
* Correspondence: weiyong@hnist.edu.cn (Y.W.); hyzhu@wzu.edu.cn (H.Z.)

Abstract

We report a passively Q-switched self-Raman laser using a dual-end composite c-cut Nd:YVO_4 crystal, which generates switchable visible emissions at 533 nm, 560 nm, and 589 nm. A Cr^{4+}:YAG/YAG composite crystal served the role of a saturable absorber to achieve passive Q-switching. An angle-tuned BBO crystal was used to achieve the frequency mixing between the first-tokes wave and the fundamental wave. At an incident pump power of 9.5 W, the maximum average output powers were 425 mW for the 589 nm yellow laser, 193 mW for the 560 nm lime laser, and 605 mW for the 533 nm green laser, with corresponding pulse widths of approximately 3.8, 3.6, and 35.1 ns, respectively. This result shows that a passive Q-switching operation with self-Raman crystals presents a promising approach for compact multi-wavelength pulse laser sources.

Keywords: visible laser; wavelength-switchable; Nd:YVO_4 laser; passively Q-switching

1. Introduction

All solid-state multi-wavelength lasers have received growing attention for their versatility in applications such as laser medical treatment [1], photodynamic therapy [2], spectral imaging [3], and other fields requiring multiple wavelength sources. For instance, retinal photocoagulation therapy requires high-power wavelength-switchable visible lasers to target different pigments in the eye [4,5]. Similarly, dermatology benefits from selective interactions with melanin, hemoglobin, or collagen, necessitating lasers with tunable wavelengths [6]. Beyond biomedical applications, these lasers are critical in defense and remote sensing, where specific wavelengths are needed for environmental monitoring, biological agent detection, and underwater communication [7–9]. In recent years, stimulated Raman scattering (SRS) has attracted widespread attention owing to its ability to facilitate flexible frequency conversion, enabling the generation of new wavelengths which are typically hard to attain with traditional solid-state lasers [10]. By combining SRS with second-harmonic generation (SHG) or sum-frequency generation (SFG), researchers have unlocked the potential for wavelength-switchable visible light emission. This approach leverages the cascaded nature of SRS, where multiple Stokes orders can be generated and subsequently frequency-doubled or mixed to produce a broad spectrum of visible and ultraviolet (UV) outputs [11].

Recent advancements in multi-wavelength visible lasers have explored both continuous-wave (CW) and pulsed regimes. For CW lasers, Chen et al. [12] and Lee et al. [13] demonstrated multi-wavelength emission by using a self-Raman laser. Among the many Raman-active crystals, a special type of neodymium (Nd)-doped crystals has attracted much attention due to their self-Raman effect. Such crystals, such as Nd:YVO_4 [14–17], Nd:$GdVO_4$ [18,19] and Nd:KGW [20,21], can simultaneously serve as laser gain media and Raman scattering media, realizing the conversion from a fundamental frequency laser to Stokes light in a single crystal, thereby greatly simplifying the laser structure. Among them, the comprehensive performance of Nd:YVO_4 is particularly outstanding. It has a high absorption cross-section and a wide absorption bandwidth at the 808 nm pump wavelength, ensuring efficient and stable pump absorption; at the same time, its high stimulated emission cross-section makes it easy to achieve low-threshold and high-gain laser operations. More importantly, the strong Raman activity of its YVO_4 crystal makes it an ideal material for building compact and efficient self-Raman lasers [22]. These advantages jointly establish the core position of Nd:YVO_4 in medium- and low-power diode pumping and integrated Raman laser technology. CW lasers have a compact structure, but their disadvantages are more obvious in applications that require high peak power [23–25]. In contrast, pulse lasers with high peak power have good applications in laser medical treatment, precision machining, and so on. Recent advances in multi-wavelength pulsed lasers primarily employ active Q-switching [26,27]. For instance, Duan et al. reported an acousto-optically Q-switched Nd:YVO_4 laser producing multiple wavelengths via self-Raman conversion, achieving visible light output with switchable wavelengths of 532, 559, 588, 620, and 657 nm at the watt power level and above [28]. However, active Q-switching systems often involve complex cavity designs and external modulators, which can limit their practicality. To address these limitations, passive Q-switching with self-Raman crystals has emerged as a promising alternative. Passive Q-switching simplifies the system architecture, enhances compactness, and reduces costs compared to active methods [29–32]. In 2022, our group achieved a 780 mW average power at 589 nm from a passively Q-switched c-cut Nd:YVO_4 self-Raman laser [33]. Despite these successes, the selective frequency mixing between fundamental and Stokes waves in passively Q-switched self-Raman lasers remains underexplored, particularly for multi-wavelength operation.

This work presents, for the first time, a triple-wavelength switchable passively Q-switched Nd:YVO_4 self-Raman laser using an angle-tuned BBO crystal. By fine-tuning the BBO crystal's orientation, we achieved SHG and SFG between the fundamental laser (1066 nm) and the first-Stokes laser (1178 nm), producing switchable outputs at 533 nm (green), 560 nm (lime), and 589 nm (yellow). Under an incident pump power of 9.5 W, the laser delivered maximum average output powers of 605, 193, and 425 mW for green, lime, and yellow emissions, respectively. At the maximum pump power, the pulse widths for yellow, lime, and green emissions were approximately 3.8, 3.6, and 35.1 ns with corresponding pulse repetition frequency of approximately 31.6, 28.3, and 34.7 kHz, respectively. These results highlight the potential of passively Q-switched self-Raman lasers as compact, efficient, and versatile sources for applications in laser medicine, display technologies, and biophotonics.

2. Crystal Analysis

As an excellent saturable absorber for passive Q-switch operations, Cr^{4+}:YAG crystal offers several advantages, including a high damage threshold, superior thermomechanical properties, and robust photochemical stability. It exhibits a low saturation energy density and high absorption cross-section around the 1 μm wavelength range, making it particularly suitable for Nd-doped solid-state lasers [34]. A Cr^{4+}:YAG/YAG composite crystal with dimensions of 3 × 3 × 5 mm^3, was employed for the passive Q-switching of the fundamental laser radiation. The principal segment is a 2 mm length Cr^{4+}:YAG crystal with an initial transmission of 85%.

The Raman gain spectrum of the YVO_4 crystal exhibits a pronounced peak at a frequency shift of 890 cm^{-1}, corresponding to a Raman gain coefficient of 4.5 cm/GW [35]. After undergoing the 890 cm^{-1} shift, the fundamental wavelength of 1066 nm in c-cut Nd:YVO_4 crystal corresponds to the first-Stokes wavelength of 1178 nm, while the fundamental wavelength of 1064 nm in an a-cut Nd:YVO_4 crystal corresponds to the first-Stokes wavelength of 1176 nm. This interaction within the Nd:YVO_4 Raman laser resonator, combined with SHG or SFG, enables tunable output across a spectral range, including green, lime, and yellow light emissions. In the experiment, a dual-end composite Nd:YVO_4 crystal, with an overall dimension of 3 × 3 × 30 mm^3, has been engineered to facilitate efficient self-Raman scattering. Both c-cut Nd:YVO_4 composite crystal and a-cut Nd:YVO_4 composite crystal were used for experimental comparison. The laser-active segment of the c-cut Nd:YVO_4 composite crystal consists of a 0.5 at.% Nd:YVO_4 crystal with dimension of 3 × 3 × 10 mm^3. The Nd:YVO_4 crystal was bonded with a 3 mm length c-cut YVO_4 cap on the pump input side to improve thermal management, while a 17 mm length c-cut YVO_4 extension was added at the output side to enhance Raman conversion efficiency. The a-cut Nd:YVO_4 composite crystal maintains the same composite structure as the c-cut version, doped with Nd^{3+} at a concentration of 0.3 at.% in the laser-active segment.

To generate three visible wavelengths via SHG and SFG involving the fundamental and first-Stokes waves, both LBO and BBO have been used as nonlinear optical crystals. The phase matching conditions required for the operation of BBO and LBO crystals are shown in Table 1. The non-critical phase matching (NCPM) LBO crystal demands temperature control over a range reaching 100 °C, which is cumbersome for practical applications. The critical phase matching (CPM) LBO crystal requires an angle adjustment of 7.7° to achieve switching for three visible wavelengths generation. And the critical phase matching (CPM) BBO crystal only requires 1.4° of angular adjustment to achieve rapid switching among the three wavelengths. Consequently, we employed an angle-tuned BBO crystal with a cut angle of (θ = 21.5°, φ = 0°) and dimensions of 4 × 4 × 7 mm^3, enabling convenient and rapid switching of visible wavelengths. The intracavity crystals were carefully wrapped in indium foil to ensure efficient thermal conductivity and then mounted on a copper cooling block, maintaining a stable temperature of approximately 20 °C.

Table 1. Phase matching parameters for the frequency mixing using LBO and BBO crystals in Nd:YVO_4 self-Raman laser.

Wavelength Conversion	1066 nm SHG	1066 nm and 1178 nm SFG	1178 nm SHG
Output wavelength/nm	533	560	589
LBO NCPM temperature/°C	146	87	40
LBO CPM angle	$\theta = 90°, \varphi = 11.2°$	$\theta = 90°, \varphi = 7.7°$	$\theta = 90°, \varphi = 3.5°$
BBO CPM angle	$\theta = 22.8°, \varphi = 0°$	$\theta = 22.1°, \varphi = 0°$	$\theta = 21.4°, \varphi = 0°$

3. Experimental Setup

Figure 1 illustrates the experimental arrangement employed in the passively Q-switched self-Raman laser, designed to produce switchable multi-wavelength output. The laser cavity consisted of a pump input mirror (IM) with a 300 mm radius of curvature, and a plane output coupler (OC) mirror. The crystals are sequentially placed inside the cavity in the order of Nd:YVO_4 composite crystal (Castech Inc., Fuzhou, China), Cr^{4+}:YAG/YAG composite crystal (Cryslaser Inc., Chengdu, China), and the BBO crystal (Castech Inc.). Those crystals had to be arranged as compactly as possible, taking into account the dimensional constraints of their respective mounts, while maintaining the total resonator length at approximately 87 mm. The pump source was a fiber-coupled (core diameter of 200 μm and numerical aperture of 0.22) laser diode operating at 808 nm. The pump beam through the coupling system consists of two lenses with focal lengths of 50 and 80 mm, and then irradiates on the Nd:YVO_4 part of composite crystal. The laser cavity supported both fundamental and Raman laser oscillation. The IM was coated with a high-reflection (HR, R > 99.9%) film for both the fundamental wavelength of 1.06 μm and the first-Stokes wavelength of 1.18 μm, while exhibiting high transmission (HT, T > 90%) at the pump wavelength of 808 nm. The OC mirror was designed with an HR coating for 1.06 μm and 1.18 μm, and an HT coating for the visible wavelength range from 530 to 600 nm. To achieve efficient visible emissions, the pump input surface of the dual-end composite Nd:YVO_4 crystal was anti-reflection (AR) coated at 808 nm, while the end surface next to the OC was HR coated in the 530–600 nm range to reverberate back visible emissions propagating backward. Additionally, both end faces of the dual-end composite Nd:YVO_4 crystal were AR coated at 1.06 μm and 1.18 μm. Similarly, AR coating was applied to both end faces of the Cr^{4+}:YAG/YAG composite crystal and BBO crystal across the wavelength ranges of 530–600 nm and 1.06–1.18 μm. The BBO crystal was fixed on a rotatable platform for angle tuning.

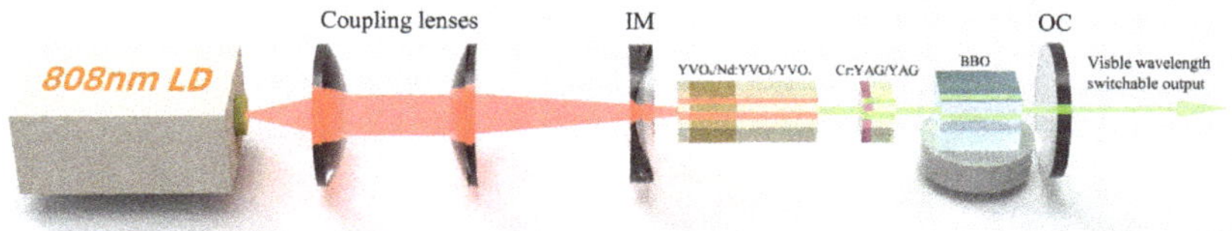

Figure 1. Experimental setup configuration of triple-wavelength switchable passively Q-switched Nd:YVO_4 self-Raman laser.

4. Experimental Results and Discussion

During the experiment, both the c-cut Nd:YVO_4 composite crystal and the a-cut Nd:YVO_4 composite crystal were utilized for experimental investigation. When a c-cut Nd:YVO_4 composite crystal was used in the laser cavity, we successfully achieved three visible wavelength switchable emissions through precise angular rotation of the BBO crystal. Using a fiber optical spectrometer (Model 3648-2-USB2, Avantes Inc., Beijing, China) with a resolution of 0.07 nm, the measured center wavelengths of the yellow–lime–green three-wavelength switchable were 589.3, 559.9, and 533.3 nm, respectively, as shown in Figure 2.

In the experiment, we optimized and investigated the output performance of the yellow, lime, and green emissions with BBO angle tuning, respectively. Figure 3 shows the visible output power against incident pump power, measured using a thermal sensor power meter (Model PM310D, Thorlabs Inc., Newton, NJ, USA). The three visible light lasers have thresholds of around 2 W. For passively Q-switched lasers, as the pump power increases, the average output power and pulse repetition rate (PRF) of the fundamental wave increase simultaneously, while the pulse energy and pulse width remain relatively constant. Consequently, the threshold pump power for the first-Stokes laser in self-Raman conversion is essentially the same as that of the fundamental wave, leading to nearly identical threshold powers for the three visible-wavelength lasers generated via their SHG and SFG. This differs significantly from the case of acousto-optic Q-switching, where the threshold of the first-Stokes laser is markedly higher than that of the fundamental wave, resulting in substantial differences among the thresholds of the three wavelengths [36]. Under identical incident pump power conditions, an approximately linear increase was observed in the average output power of the visible lasers. The slope efficiencies derived from linear fitting were 5.8%, 2.7%, and 7.6%, respectively. When the incident pump power reached 9.5 W, the average output powers were 425 mW, 193 mW, and 605 mW for yellow, lime, and green emissions, with respective diode-to-visible conversion efficiencies of 4.5%, 2%, and 6.3%, respectively. Among these, the 533 nm wavelength laser demonstrated the highest average output power performance.

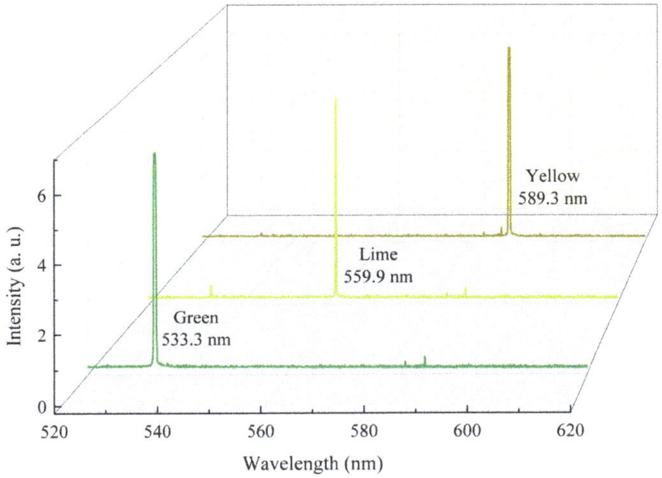

Figure 2. Measured laser spectra of three visible lasers.

In addition, the average output power and efficiency of SFG are significantly lower than those of SHG. The main reason is that Q-switched Raman lasers exhibit pulse compression characteristics. The primary reason for this discrepancy lies in the temporal non-overlap between the fundamental laser and in the fact that the first-Stokes laser pulses significantly reduce the SFG conversion efficiency. The pulse waveforms of the three output wavelengths were measured using InGaAs free-space photodetectors (Model DETO8C/M, Thorlabs Inc., Newton, NJ, USA) and shown on a 500 MHz digital oscilloscope (Model DPO3052B, Tektronix Inc., Beaverton, OR, USA), as illustrated in Figure 4. For the green laser at 533 nm, lime laser at 560 nm, and yellow laser at 589 nm, the measured pulse widths were approximately 35.1, 3.6, and 3.8 ns, respectively. Notably, the SHG pulse width for the fundamental laser was substantially broader compared to both the SHG pulse width of the first-Stokes laser and the SFG pulse width produced by the fundamental and first-Stokes light. The pulse width characteristics of the SFG output further reflect the temporal relationship between the fundamental and first-Stokes laser pulses. It can be concluded that the broader pulse width of the fundamental light compared to the first-Stokes light results in a lower temporal overlap, thereby reducing the SFG conversion efficiency in Q-switched Raman lasers.

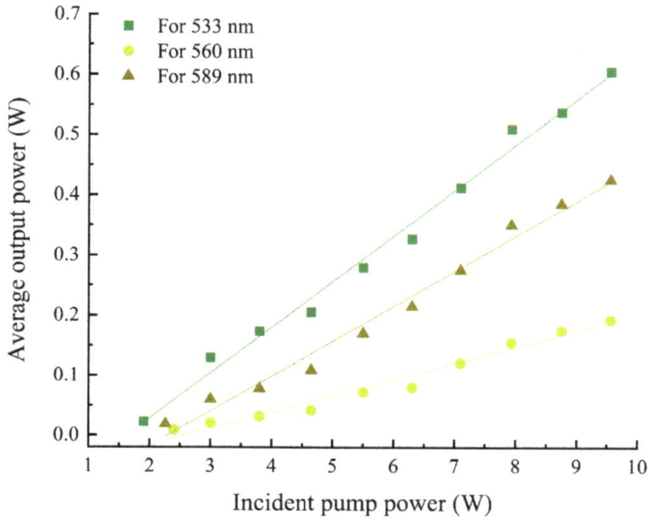

Figure 3. Average output power characteristics of three visible lasers (yellow, lime and green) versus incident pump power.

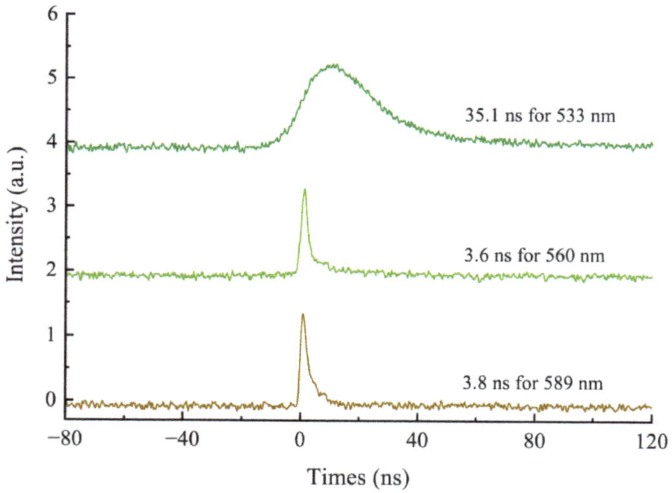

Figure 4. Measured pulse widths for green (533 nm), lime (560 nm), and yellow (589 nm) lasers at the incident pump power of 9.5 W.

The train of pulses for each of the three visible-wavelength lasers are shown in Figure 5. The pulse repetition rate of three visible-wavelength lasers are related to the power density of the intracavity fundamental laser. It is precisely the varying losses imposed on the fundamental laser during the frequency conversion process that result in differences in their corresponding pulse repetition rate. The measured PRF of 533, 560, and 589 nm lasers were about 34.7, 28.3, and 31.6 kHz, respectively, for the pump power of 9.5 W. Because the fundamental and the first-Stokes wave do not overlap, conversion efficiency is reduced and pulse peak-to-peak stability worsens. Consequently, the 560 nm emission has the worst peak-to-peak stability.

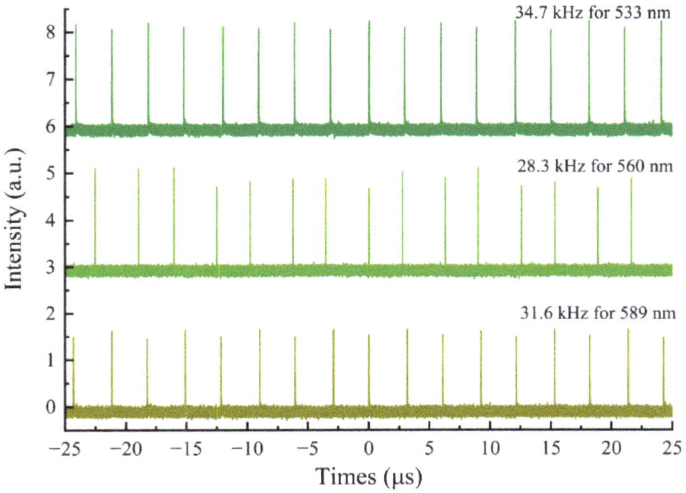

Figure 5. Measured pulse trains for green (533 nm), lime (560 nm), and yellow (589 nm) lasers at the incident pump power of 9.5 W.

Further, by replacing c-cut with a-cut Nd:YVO_4 composite crystals, we plan to study the multi-wavelength laser output characteristics using a-cut Nd:YVO_4 composite crystal and its differences from c-cut Nd:YVO_4 composite crystals. When using an a-cut Nd: YVO_4 composite crystal with the similar laser system, only obvious green light was observed during the experiment, reaching an average output power of 420 mW when the pump power was 9.5 W. However, by rotating the BBO angle, no other colors of light appeared, indicating that Raman conversion under passive Q-switching was not achieved. This issue may be due to the fact that the a-cut Nd: YVO_4 crystal has a larger stimulated emission cross-section ($\sigma_{\text{a-cut}} = 25 \times 10^{-19}$ cm^2), which results in insufficient pulse peak power of the fundamental wave to initiate Raman conversion. And c-cut Nd: YVO_4 exhibits a lower stimulated emission cross-section ($\sigma_{\text{c-cut}} = 6.5 \times 10^{-19}$ cm^2), which promotes shorter pulse duration and elevated peak power in Q-switching [37,38]. In 2004, Chen also discovered that a-cut Nd:YVO_4 could not achieve Raman conversion during his research on Nd: YVO_4 self-Raman lasers with passive Q-switching [39].

The significance of this study lies in demonstrating a simplified and compact multi-wavelength laser system that integrates passive Q-switching with self-Raman conversion. By eliminating active modulation requirements and leveraging the intrinsic properties of c-cut Nd:YVO_4 crystals, this approach provides a practical tunable visible laser source with high peak power. Although selective frequency mixing has not yet been achieved in the self-Raman configuration using a-cut Nd:YVO_4 crystal, future work should focus on increasing the fundamental wave power density and peak power, which could be achieved by optimizing the cavity structure (e.g., adopting a folded cavity) and the initial transmission of the saturable absorber. These enhancements would increase fundamental wave power density and peak power, potentially enabling improvement of Raman conversion efficiency and achievement of selective frequency mixing in the self-Raman configuration using an a-cut Nd:YVO_4 crystal. Further research directions include optimizing pulse stability and efficiency, as well as extending the operational wavelength range for specialized applications.

5. Conclusions

In summary, we first proposed a triple-wavelength switchable passively Q-switched self-Raman laser system with c-cut Nd:YVO_4 crystals. By fine-tuning the BBO crystal's angle, we successfully achieved SHG and SFG between the fundamental laser and the first-Stokes laser, enabling wavelength-switchable outputs at 533 nm (green), 560 nm (lime), and 589 nm (yellow). With the incident pump power of 9.5 W, the maximum average output powers measured were 425 mW at 589 nm, 193 mW at 560 nm, and 605 mW at 533 nm., with respective pulse widths of approximately 3.8, 3.6, and 35.1 ns, respectively. Under maximum pump power conditions, the pulse repetition frequencies were approximately 31.6, 28.3, and 34.7 kHz, while the single-pulse energies were 13.4, 6.8, and 17.4 μJ for the yellow, lime, and green emissions, respectively. We also found that a-cut Nd:YVO_4 is difficult to achieve passive Q-switching self-Raman conversion with similar laser setup. These results indicate that the passively Q-switched self-Raman multi-wavelength switchable laser using a c-cut Nd:YVO_4 composite crystal is compact, easy to implement, and has broad prospects. This compact visible wavelength switchable laser has possible applications in fields like laser medicine, display, and biophotonics.

Author Contributions: Conceptualization, S.L., Y.D. and H.Z.; methodology, S.L., S.Z. and H.Z.; software, S.Z.; validation, S.L., S.Z. and B.Z.; formal analysis, S.L., S.Z., Y.W. and Y.Z.; investigation, S.L., S.Z., and B.Z.; resources, Y.D. and H.Z.; data curation, S.L., Y.Z., Y.D. and H.Z.; writing—original draft preparation, S.L., S.Z.; writing—review and editing, Y.W. and H.Z.; supervision, H.Z.; project administration, Y.W. and H.Z.; funding acquisition, Y.W., Y.D. and H.Z. All authors have read and agreed to the published version of the manuscript.

Funding: This research was funded by the National Natural Science Foundation of China, grant number 62275200 and 62075167, and the Natural Science Foundation of Zhejiang Province, grant number LQN25F050012, and the Natural Science Foundation of Hunan Province, grant number 2025JJ70264, and the Basic scientific research project of Wenzhou City, grant number L20240003.

Institutional Review Board Statement: Not applicable.

Informed Consent Statement: Not applicable.

Data Availability Statement: All of the data reported in the paper are presented in the main text. Any other data will be provided on request.

Conflicts of Interest: The authors declare no conflicts of interest.

References

1. Gu, Y.G.; Niu, J.; Yang, J.; Xu, H.X. Application of laser in the medical field. *Chin. Opt.* **2023**, *16*, 283–295.
2. Kachynski, A.V.; Pliss, A.; Kuzmin, A.N.; Ohulchanskyy, T.Y.; Baev, A.; Qu, J.; Prasad, P.N. Photodynamic therapy by in situ nonlinear photon conversion. *Nat. Photonics* **2014**, *8*, 455–461. [CrossRef]
3. Pask, H.M.; Dekker, P.; Mildren, R.P.; Spence, D.J.; Piper, J.A. Wavelength-versatile visible and UV sources based on crystalline Raman lasers. *Prog. Quantum Electron.* **2008**, *32*, 121–158. [CrossRef]
4. Kozak, I.; Luttrull, J.K. Modern retinal laser therapy. *Saudi J. Ophthalmol.* **2015**, *29*, 137–146. [CrossRef]
5. Lock, J.H.J.; Fong, K.C.S. An update on retinal laser therapy. *Clin. Exp. Optom.* **2011**, *94*, 43–51. [CrossRef]
6. Alexiades-Armenakas, M.R.; Dover, J.S.; Arndt, K.A. The spectrum of laser skin resurfacing: Nonablative, fractional, and ablative laser resurfacing. *J. Am. Acad. Dermatol.* **2008**, *58*, 719–737. [CrossRef]
7. Hammann, E.; Behrendt, A.; Le Mounier, F.; Wulfmeyer, V. Temperature profiling of the atmospheric boundary layer with rotational Raman lidar during the HD(CP)^2 Observational Prototype Experiment. *Atmos. Chem. Phys.* **2015**, *15*, 2867–2881. [CrossRef]
8. Li, X.; Huang, S.H.; Sun, Z.H. Technology and equipment development in laser-induced fluorescence-based remote and field detection of biological aerosols. *J. Biosaf. Biosecur.* **2019**, *1*, 113–122. [CrossRef]
9. Zeng, Z.Q.; Fu, S.; Zhang, H.H.; Dong, Y.H.; Cheng, J.L. A Survey of Underwater Optical Wireless Communications. *IEEE Commun. Surv. Tutor.* **2017**, *19*, 204–238. [CrossRef]

10. Huo, X.W.; Qi, Y.Y.; Zhang, Y.; Chen, B.; Bai, Z.X.; Ding, J.; Wang, Y.L.; Lu, Z.W. Research development of 589 nm laser for sodium laser guide stars. *Opt. Lasers Eng.* **2020**, *134*, 106207. [CrossRef]
11. Chen, H.; Cui, Y.F.; Li, X.W.; Zhang, B.Y.; Cai, Y.P.; Ding, J.; Qi, Y.Y.; Yan, B.Z.; Wang, Y.L.; Lu, Z.W.; et al. High-power dual-wavelength intracavity diamond Raman laser. *Funct. Diam.* **2023**, *3*, 2282527. [CrossRef]
12. Chen, Y.F.; Pan, Y.Y.; Liu, Y.C.; Cheng, H.P.; Tsou, C.H.; Liang, H.C. Efficient high-power continuous-wave lasers at green-lime-yellow wavelengths by using a Nd:YVO_4 self-Raman crystal. *Opt. Express* **2019**, *27*, 2029–2035. [CrossRef] [PubMed]
13. Lee, A.J.; Spence, D.J.; Piper, J.A.; Pask, H.M. A wavelength-versatile, continuous-wave, self-Raman solid-state laser operating in the visible. *Opt. Express* **2010**, *18*, 20013–20018. [CrossRef]
14. Ding, S.H.; Zhang, X.Y.; Wang, Q.P.; Su, F.F.; Jia, P.; Li, S.T.; Fan, S.Z.; Chang, J.; Zhang, S.S.; Liu, Z.J. Theoretical and experimental study on the self-Raman laser with Nd:YVO_4 crystal. *IEEE J. Quantum Electron.* **2006**, *42*, 927–933. [CrossRef]
15. Fan, L.; Zhao, W.Q.; Qiao, X.; Xia, C.Q.; Wang, L.C.; Fan, H.B.; Shen, M.Y. An efficient continuous-wave YVO_4/Nd: YVO_4/YVO_4 self-Raman laser pumped by a wavelength-locked 878.9 nm laser diode. *Chin. Phys. B* **2016**, *25*, 114207. [CrossRef]
16. She, K.; Zhou, P.F.; Wei, Y.; Xu, S.; Li, B.X.; Liao, W.B.; Lin, Z.L.; Zhang, G. Manipulation of linearly polarized states in a c-cut Nd: YVO_4 passively Q-switched self-Raman laser. *Appl. Phys. B* **2024**, *130*, 51. [CrossRef]
17. Wang, X.L.; Wang, X.J.; Dong, J. Multiwavelength, sub-nanosecond Yb: YAG/Cr^{4+}: YAG/YVO_4 passively Q-switched Raman microchip laser. *IEEE J. Sel. Top. Quantum Electron.* **2018**, *24*, 1–8.
18. Ma, Y.; Lee, A.J.; Pask, H.M.; Miyamoto, K.; Omatsu, T. Direct generation of 1108 nm and 1173 nm Laguerre-Gaussian modes from a self-Raman Nd:$GdVO_4$ laser. *Opt. Express* **2020**, *28*, 24095–24103. [CrossRef]
19. Chen, Y.F.; Huang, H.Y.; Lee, C.C.; Hsiao, J.Q.; Tsou, C.H.; Liang, H.C. High-power diode-pumped Nd:$GdVO_4$/KGW Raman laser at 578 nm. *Opt. Lett.* **2020**, *45*, 5562–5565. [CrossRef]
20. Zou, Q.S.; Sun, Q.H.; Dai, Z.H.; Ding, S.H. Efficient passively Q-switched Nd: KGW/Cr^{4+}:YAG self-Raman laser. *Appl. Phys. B* **2024**, *130*, 150. [CrossRef]
21. Tang, C.Y.; Zhuang, W.Z.; Su, K.W.; Chen, Y.F. Efficient Continuous-Wave Self-Raman Nd:KGW Laser With Intracavity Cascade Emission Based on Shift of 89 cm^{-1}. *IEEE J. Sel. Top. Quantum Electron.* **2015**, *21*, 142–147. [CrossRef]
22. Koechner, W. *Solid-State Laser Engineering*, 6th ed.; Springer: New York, NY, USA, 2006.
23. Hsieh, C.L.; Huang, H.J.; Chen, C.L.; Liang, H.V.; Chen, Y.F. Selectable two-wavelength Nd:YVO_4 Raman laser at 671 and 714 nm. *Opt. Lett.* **2023**, *48*, 1510–1513. [CrossRef]
24. Lee, C.Y.; Chang, C.C.; Sung, C.L.; Chen, Y.F. Intracavity continuous-wave multiple stimulated-Raman-scattering emissions in a KTP crystal pumped by a Nd:YVO_4 laser. *Opt. Express* **2015**, *23*, 22765–22770. [CrossRef]
25. Li, D.; Huang, C.Y.; Chang, X.W.; Liang, H.C.; Chen, Y.F. Continuous-Wave Self-Raman Vanadate Lasers Generating Versatile Visible Wavelengths. *Photonics* **2024**, *11*, 601. [CrossRef]
26. Chen, Z.P.; Chen, Z.H.; Pan, H.G.; Zhang, A.L.; Zhang, B.; Li, R.P.; Chen, C.Q.; Peng, Y.M.; Liu, Z.Z. Tunable and switchable multi-wavelength actively Q-switched erbium-doped fiber laser based on cascaded fiber filters and EOM. *J. Opt. India* **2023**, *52*, 2333–2341. [CrossRef]
27. Zhang, H.N.; Chen, X.H.; Wang, Q.P.; Li, P. Dual-wavelength actively Q-switched diode-end-pumped ceramic Nd:YAG/$BaWO_4$ Raman laser operating at 1240 and 1376 nm. *Laser Phys. Lett.* **2014**, *11*, 105806. [CrossRef]
28. Duan, Y.M.; Zhu, H.Y.; Zhang, G.; Tang, D.Y. Multiple visible wavelength switchable cascaded self-Raman laser based on selective wave-mixing mechanism. *Appl. Phys. Lett.* **2023**, *123*, 261102. [CrossRef]
29. Ma, Y.F.; Li, X.D.; Yu, X.; Fan, R.W.; Yan, R.P.; Peng, J.B.; Xu, X.R.; Sun, R.; Chen, D.Y. A novel miniaturized passively Q-switched pulse-burst laser for engine ignition. *Opt. Express* **2014**, *22*, 24655–24665. [CrossRef] [PubMed]
30. Mao, W.J.; Zhang, D.; Lu, H.Q.; Zhu, X.L.; Li, Z.H.; Wang, H.Y.; Duan, Y.M.; Zhu, H.Y. Compact passively Q-switched KTA self-frequency doubled Raman laser with 671 cm^{-1} shift. *Opt. Laser Technol.* **2022**, *156*, 108619. [CrossRef]
31. Tang, J.L.; Bai, Z.X.; Zhang, D.; Qi, Y.Y.; Ding, J.; Wang, Y.L.; Lu, Z.W. Advances in All-Solid-State Passively Q-Switched Lasers Based on Cr^{4+}:YAG Saturable Absorber. *Photonics* **2021**, *8*, 93. [CrossRef]
32. Zhang, L.; Duan, Y.M.; Sun, Y.L.; Chen, Y.J.; Li, Z.H.; Zhu, H.Y.; Zhang, G.; Tang, D.Y. Passively Q-switched multiple visible wavelengths switchable YVO_4 Raman laser. *J. Lumin.* **2020**, *228*, 117650. [CrossRef]
33. Li, Y.H.; Huang, X.H.; Mao, W.J.; Xu, J.; Duan, Y.M.; Zhu, H.Y. Compact 589 nm yellow source generated by frequency-doubling of passively Q-switched Nd:YVO_4 Raman laser. *Microw. Opt. Technol. Lett.* **2023**, *65*, 1122–1126. [CrossRef]
34. Ding, S.L.; Yang, X.T.; Zhang, Q.L.; Liu, W.P.; Luo, J.Q.; Sun, G.H.; Ma, Y.F.; Sun, D.L. Diode-pumped passively Q-switched Nd : $GdNbO_4$ laser with Cr^{4+} : YAG saturable absorber. *Opt. Eng.* **2017**, *56*, 086111. [CrossRef]
35. Kaminskii, A.A.; Ueda, K.; Eichler, H.J.; Kuwano, Y.; Kouta, H.; Bagaev, S.N.; Chyba, T.H.; Barnes, J.C.; Gad, G.M.; Murai, T. Tetragonal vanadates YVO_4 and $GdVO_4$–new efficient $\chi^{(3)}$-materials for Raman lasers. *Opt. Commun.* **2001**, *194*, 201–206. [CrossRef]
36. Sun, Y.L.; Duan, Y.M.; Cheng, M.Y.; Yuan, X.Z.; Zhang, L.; Zhang, D.; Zhu, H.Y. Triple wavelength-switchable lasing in yellow-green based on frequency mixing of self-Raman operation. *Acta Phys. Sin.* **2020**, *69*, 124201. [CrossRef]

37. Chen, H.; Wu, E.; Zeng, H.P. Comparison between α-cut and off-axially cut Nd:YVO_4 lasers passively Q-switched with a Cr^{4+}:YAG crystal. *Opt. Commun.* **2004**, *230*, 175–180. [CrossRef]
38. Duan, Y.M.; Xu, J.; Wei, Y.; Jin, X.X.; Li, Z.H.; Zhu, H.Y. Yellow-orange wavelength-switchable laser emission generated from c-cut Nd:YVO_4 self-Raman with 890 and 259 cm^{-1} shifts. *J. Lumin.* **2024**, *267*, 120402. [CrossRef]
39. Chen, Y.F. Efficient subnanosecond diode-pumped passively Q-switched Nd:YVO_4 self-stimulated Raman laser. *Opt. Lett.* **2004**, *29*, 1251–1253. [CrossRef]

photonics

MDPI

Article

Miniaturized Multicolor Femtosecond Laser Based on Quartz-Encapsulated Nonlinear Frequency Conversion

Bosong Yu [1], Siying Wang [2], Aimin Wang [2,3], Yizhou Liu [4,5,*] and Lishuang Feng [1,6]

1 School of Instrumentation and Optoelectronic Engineering, Beihang University, Beijing 100191, China; bosongyu@buaa.edu.cn (B.Y.); fenglishuang@buaa.edu.cn (L.F.)
2 School of Electronics, Peking University, Beijing 100871, China; siyingwang@pku.edu.cn (S.W.); wangaimin@pku.edu.cn (A.W.)
3 State Key Laboratory of Advanced Optical Communication System and Networks, School of Electronics, Peking University, Beijing 100871, China
4 School of Information Science and Engineering, Shandong Provincial Key Laboratory of Laser Technology and Application, Shandong University, Qingdao 266237, China
5 Key Laboratory of Laser & Infrared System Ministry of Education, Shandong University, Qingdao 266237, China
6 Key Laboratory of Precision Opto-Mechatronics Technology (Ministry of Education), Beihang University, Beijing 100191, China
* Correspondence: yizhou.liu@sdu.edu.cn

Abstract

Ultrafast lasers operating at 740 nm and 820 nm have attracted widespread attention as two-photon light sources for the detection of biological metabolism. Here, we report on a solid-like quartz-encapsulated femtosecond laser with a repetition rate of 80 MHz, delivering 740 nm and 820 nm femtosecond laser pulses. This home-built laser system was realized by employing an erbium-doped 1560 nm fiber laser as the fundamental laser source. A quartz-encapsulated nonlinear frequency conversion stage, consisting of a second-harmonic generation (SHG) stage and self-phase modulation (SPM)-based nonlinear spectral broadening stage, was utilized to deliver 30 mW, 53.7 fs, 740 nm laser pulses and the 15 mW, 60.8 fs, 820 nm laser pulses. Further imaging capabilities of both wavelengths were validated using a custom-built inverted two-photon microscope. Clear imaging results were obtained from mouse kidney sections and pollen samples by collecting the corresponding fluorescence signals. The achieved results demonstrate the great potential of this laser source for advanced two-photon microscopy in metabolic detection.

Keywords: solid-like femtosecond laser; nonlinear frequency conversion; two-photon microscopy

1. Introduction

Two-photon microscopy (TPM), widely recognized as one of the most effective optical techniques for real-time, in vivo, and non-invasive imaging of live tissues and biological systems, has found broad applications across diverse biomedical fields, such as brain imaging [1,2], endoscopy [3,4], intravital pathology [5,6], and dermatological diagnostics [7, 8]. TPM induces two-photon absorption of endogenous fluorophores within cells, enabling label-free imaging of cellular structures and metabolic activity using femtosecond excitation wavelengths between 700 and 900 nm [9,10]. In particular, two-photon excitation of the intrinsic cofactors flavin adenine dinucleotide (FAD, efficiently excited at 820 nm) and nicotinamide adenine dinucleotide (NADH, efficiently excited at 740 nm) allows for simultaneous acquisition of both structural and metabolic information of biological tissues [11,12], which is especially valuable for applications such as cancer diagnostics [13] and metabolic profiling [14]. Therefore, the femtosecond lasers operating at both 740 nm and 820 nm can efficiently excite these endogenous fluorophores and improve the accuracy and efficiency of the non-invasive in vivo cellular detection [15].

For the past decades, femtosecond Ti:sapphire lasers have been the generally utilized light sources for two-photon imaging systems, delivering <100 fs pulses with the wavelength range spanning from 700 nm to 1100 nm [16,17]. However, the complex systematic construction and high costs of Ti:sapphire lasers limit their further practical applications in advanced biomedical research [18]. Fiber lasers have attracted increasing attention due to their structural stability and simple systematic design [19]. L. Huang et al. demonstrated a 61 mW, 80 fs, 790 nm laser source employing the SHG with an Er-doped 1560 nm fiber laser, subsequently utilized to realize the imaging of the femurs of mice and pigs [11]. L.T. Chou et al. employed a 48 MHz, 120 fs, ytterbium-doped, 1 μm fiber laser to produce 740 nm pulses with an average power of 91.2 mW with the SPM [20]. This 740 nm femtosecond laser was further utilized to illuminate mouse MCF-7 breast tumor cells inside the TPM. The complexity of fiber laser systems for autofluorescence applications, which leads to long-term instability, hinders their practical use. Addressing these issues, the SHG of the 1560 nm nonlinearly amplified femtosecond laser and the SPM of the frequency-doubled 780 nm femtosecond laser can be quartz-encapsulated together in generating 740 nm and 820 nm femtosecond lasers. To our knowledge, this kind of femtosecond laser delivering both 740 nm and 820 nm laser contents is rarely reported. Furthermore, this design features high power stability and a compact packaging structure, enabling a plug-and-play connection between the laser source and the imaging probe. Therefore, an efficient two-photon excitation can be achieved, facilitating the acquisition of both morphological and metabolic information from tissues and cells.

In this study, we report on a solid-like, quartz-encapsulated femtosecond laser source with a repetition rate of 80 MHz, delivering dual-wavelength pulses at 740 nm and 820 nm. The system originates from a fully polarization maintenance (PM) Er-doped fiber laser operating at 1560 nm, which produces 52.4 fs pulses. These fundamental pulses are first frequency-doubled to 780 nm in a periodically poled lithium niobate (PPLN) crystal. The resulting light is then coupled into a photonic crystal fiber (NL-PM750) for spectral broadening via supercontinuum generation. To ensure exceptional long-term stability, all nonlinear optical components are integrated and sealed within a robust quartz module. This compact and highly reliable source simultaneously provides two distinct outputs: 53.7 fs pulses at 740 nm with an average power of 30 mW and 60.8 fs pulses at 820 nm with an average power of 15 mW, making it highly suitable for applications such as two-photon microscopy.

2. Laser Setup and Quartz-Encapsulated Structure

Figure 1a illustrates the experimental configuration of the frequency-doubled Er-doped fiber laser system. The seed laser is an 80 MHz mode-locked Er-doped fiber oscillator utilizing a nonlinear amplifying loop mirror (NALM), delivering pulses with a duration of 55 fs [21,22]. A 4.6 m PM fiber (PM1550, Coherent, Saxonburg, PA, USA) was employed to pre-manage the optical chirp of the seed laser pulses. The all-PM fiber amplifier consisted of a PM wavelength division multiplexer, two 850 mW 976 nm single-mode laser diodes, and a 3.6 m Er-doped gain fiber (DHB1500, Fibercore, Southampton, UK). It was utilized to scale up the average power of the seed laser from 2.9 mW to 350 mW. The combination of the Er-doped gain fiber (29.2 fs^2/mm) and the PM1550 (-23.5 fs^2/mm) spliced before and after the gain fiber was utilized to enable the nonlinear amplification and compression processes [23]. Based on contributions of the dispersion management and the self-phase modulation, the optical pulse duration of the nonlinearly amplified 357 mW 1560 nm laser pulses was nonlinearly compressed to 52.4 fs based on the Gaussian assumption, directly delivered from the fiber output port.

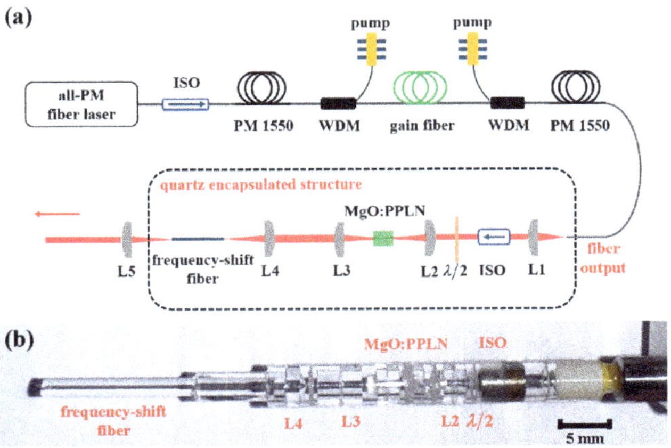

Figure 1. (**a**) Experimental configuration of the Er-doped fiber laser system. (**b**) Optical element quartz-encapsulated structure.

The 357 mW, 80 MHz laser pulses were further delivered into a quartz-encapsulated module to realize the optical collimation, optical isolation, frequency doubling, polarization management, and nonlinear spectral broadening. Figure 1b illustrates the real picture of the quartz-encapsulated module. The 357 mW, 1560 nm laser beam was firstly collimated with an aspheric lens L1 (f = 11 mm, 354064C, Lightpath, Orlando, FL, USA). The corresponding collimated beam diameter was 2.7 mm. A combination of a half waveplate, a focusing aspheric lens L2 (f = 6.2 mm, 354171C, LightPath, Orlando, FL, USA), a MgO:PPLN (C-M-1940-S-1, Foctek Photonics, Fuzhou, China) nonlinear crystal, and a collimating aspheric lens L3 (f = 6.2 mm, 354171B, LightPath, Orlando, FL, USA) was utilized to realize the SHG of the 1560 nm femtosecond laser. The poling period of 19.7 μm of the 0.5 mm long MgO:PPLN crystal was employed in realizing the nonlinear phase matching condition. The optimal quasi-phase-matching temperature of the crystal was designed to be room temperature, eliminating additional active thermal management. The generated 148 mW, 80 MHz, 780 nm frequency-doubled laser pulses were further coupled into a high nonlinearity photonic crystal fiber (HNPCF, NL-PM-750, NKT Photonics, Blokken, Birkerød, Denmark) with an aspheric lens L4 (f = 1.45 mm, 354140B, LightPath, Orlando, FL, USA) to realize the nonlinear spectral broadening. The NL-PM-750 photonic crystal fiber with a tightly confined mode-field diameter of approximately 1.6 μm, featuring near-zero dispersion at 750 nm and a group velocity dispersion (GVD) parameter of $\beta_2 = -0.0066\ \text{ps}^2/\text{m}$ at 780 nm, enabled the efficient SPM-induced nonlinear spectral broadening process over 2 cm HNPCF. The realized coupling efficiency was measured to be approximately 70% with a coupled pulse energy of around 1.25 nJ, indicating a perfect optical coupling construction.

Realizing the quartz-encapsulated module, the utilized aspheric lenses, waveplates, free-space isolator, and nonlinear crystal were carefully encapsulated inside the quartz glass tubes with identical outer diameters. To achieve structural integration and stable coupling, the HNPCF was firstly collapsed at the input port and then inserted into a quartz glass sleeve for further protection. A quartz glass tube with the same outer diameter as the fiber's quartz sleeve was subsequently bonded to the rear side of the coupling lens (L4), ensuring that the fiber end face was precisely positioned at the focal plane of the lens. To further enhance the alignment accuracy and bonding strength between the photonic crystal fiber (encapsulated in the quartz sleeve) and the coupling lens, an additional quartz glass tube was employed to connect the two components, thereby reinforcing mechanical stability and enabling integrated assembly. All optical elements were precisely aligned and fixed using UV-curable optical adhesives. Specifically, the end-face bonding of quartz tubes was performed using EM10483 (EMIUV, Breckenridge, CO, USA), a high-glass transition temperature adhesive that offers excellent thermal and mechanical stability. For the interface between the optical components and the inner wall of the quartz housing, NOA61 (Norland, Jamesburg, NJ, USA) was selected for its superior fluidity, which promotes better wetting and increases the effective bonding area. UV curing was conducted under controlled conditions to avoid thermal stress and ensure stable alignment. This hybrid bonding strategy enabled the rigid, low-shrinkage encapsulation of all components, suppressing vibration- and temperature-induced misalignments. The overall packaging approach ensured precise optical alignment and robust mechanical integration of the SHG and frequency-shift modules, thereby enhancing the long-term stability and reliability of the output femtosecond pulses.

3. Results

Figure 2a illustrates the optical spectrum of the nonlinearly compressed 347 mW, 80 MHz, 1560 nm signal pulses delivered by the output port of the PM1550 compression fiber. The measured 3 dB spectral bandwidth and 10 dB bandwidth were 3 nm and 64 nm, respectively. Figure 2b presents the measured autocorrelation trace of the compressed pulses. The corresponding pulse duration was estimated to be 52.4 $\times$ 1.414 fs based on the Gaussian assumption. Notably, the trace exhibits pronounced pedestals, which originate from uncompensated high-order phase components—primarily attributed to residual third-order dispersion. Figure 2c plots the measured optical pulse train with the repetition rate of 80 MHz, indicating a stable mode-locking operation state of the 1560 nm fiber oscillator. Figure 2d plots the measured root-mean-square (RMS) of 0.19% @1h of the delivered 347 mW nonlinearly amplified 1560 nm laser pulses under the coupled pump power of 1.7 W. The output power was monitored with a high-resolution power meter (S314C) (Thorlabs, Newton, MA, USA) with a resolution of 5 μW. Through careful dispersion and nonlinear phase management, a femtosecond fiber laser output was ultimately achieved, delivering 55 fs pulses at an 80 MHz repetition rate with an average power of 347 mW. This output served as the pump source for subsequent nonlinear frequency conversion processes.

Figure 3a illustrates the measured optical spectrum of the SHG pulses, centered at 780 nm with a full width at half maximum (FWHM) of 7 nm. Figure 3b shows the corresponding autocorrelation trace, and the measured pulse duration was estimated to be 95 $\times$ 1.414 fs, assuming a Gaussian profile. As shown in Figure 3d, the calculation of the phase-matching condition reveals that the optimal poling period for the PPLN crystal is 19.7 μm at a temperature of 300 K. The frequency doubling process yielded an average output power of 145 mW, corresponding to a conversion efficiency of 40.3%. Through appropriate selection and design, the quasi-phase-matching temperature of the frequency-doubling crystal was set to room temperature. In the context of biological imaging applications, the combination of low average power and the absence of active thermal control was identified as one of the contributing factors to the high stability of the system. A third aspheric lens (354171B, LightPath, Orlando, FL, USA) was employed after the PPLN crystal to collimate the frequency-doubled output beam. The resulting output beam profile is shown in Figure 3c, exhibiting a beam diameter of 1.45 mm and an ellipticity of 0.95, indicating good spatial quality. Furthermore, one-hour power stabil Orlando, USA ity measurements were performed. The RMS of the power fluctuation was measured to be 0.17%, demonstrating excellent long-term stability of the SHG femtosecond output.

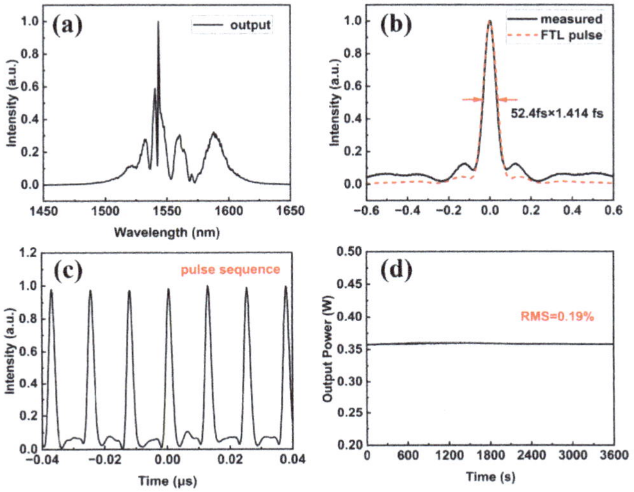

Figure 2. Output characteristics of the 1560 nm Er-doped fiber laser. (**a**) Optical spectrum of the nonlinearly amplified 1560 nm laser. (**b**) Autocorrelation trace of the nonlinearly compressed 1560 nm laser pulses with the pulse duration of 52.4 × 1.414 fs based on the Gaussian assumption (black solid line) and the calculated Fourier-transform limited autocorrelation trace (red dashed line). (**c**) Mode-locking pulse train with the repetition rate of 80 MHz. (**d**) Measured optical power stability over one hour.

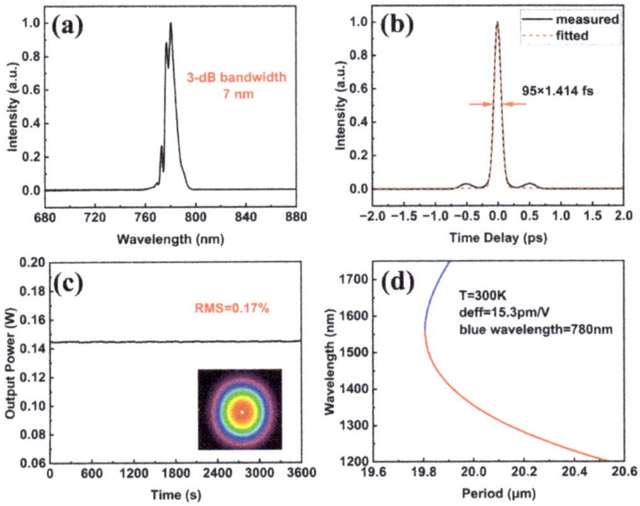

Figure 3. Characterization of the frequency-doubled pulses generated with the MgO:PPLN crystal. (**a**) Optical spectrum centered at 780 nm. (**b**) Autocorrelation trace of the 780 nm femtosecond pulse (black solid line) and the Gaussian fit (red dashed line). (**c**) Long-term optical power stability of the output power (top) and the corresponding beam profile (bottom). (**d**) Calculated phase-matching curve for the MgO:PPLN crystal at 300 K.

Figure 4a illustrates the dispersion and loss parameters of the HNPCF; the near-zero dispersion point of this HNPCF is around 750 nm. Figure 4b shows the output spectrum after nonlinear broadening. A distinct dual-peak structure is observed, with the left peak centered at 740 nm and the right peak centered at 820 nm. The short-wavelength peak arises primarily from blue-shifted frequency components at the leading edge of the pulse, while the long-wavelength peak originates from red-shifted components at the trailing edge.

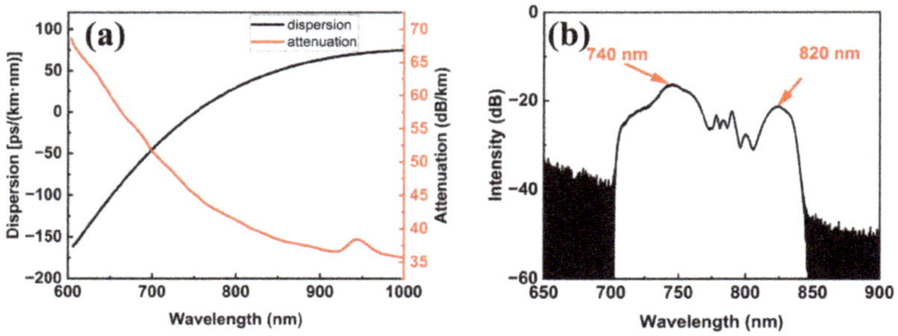

Figure 4. Nonlinear spectral broadening in the HNPCF. (**a**) Group velocity dispersion (black solid line) and confinement loss (red solid line) of the HNPCF. (**b**) Output optical spectrum after the nonlinear spectral broadening process.

After collimation of the output beam using an aspheric lens L5 (354140B, LightPath, Orlando, FL, USA), spectral filtering was performed using a 760 nm short-pass filter (SP760, JiuShi Optics, Shanghai, China). The resulting output pulse exhibited an average power of 30 mW. The corresponding filtered spectrum is shown in Figure 5a, with a central wavelength of 741 nm. The autocorrelation trace of the output pulse is presented in Figure 5b, and the pulse duration, assuming a Gaussian temporal profile, was measured to be 53.7 × 1.414 fs. Subsequently, spectral filtering was also carried out using an 800 nm long-pass filter (LP800, JiuShi Optics, Shanghai, China). The filtered output exhibited an average pulse power of 15 mW. The corresponding spectrum, with a central wavelength of 820 nm, is depicted in Figure 5c. The autocorrelation trace of this signal is shown in Figure 5d, yielding a pulse duration of 60.8 × 1.414 fs under the Gaussian assumption.

The 740 nm and 820 nm femtosecond pulses were further routed to a miniaturized multiphoton microscopy probe to evaluate its imaging performance. The imaging setup is shown in Figure 6a. In this setup, the excitation light path is marked in red and the collection light path is marked in green. After passing through the filter, the collimated beam was directly coupled into the probe without the need for anti-resonant fibers or photonic crystal fibers. The inverted two-photon microscope used for imaging validation consists of a Micro-Electro-Mechanical System (MEMS) scanning mirror, a scan lens, a dichroic mirror, an objective lens, and a condenser lens, as shown in Figure 6a. The MEMS scanner operates at a frequency of 2400 Hz, with a frame rate of 3 fps and a field of view of 300 μm × 300 μm. The excitation light is focused by a 9× microscope objective (N.A. = 0.7). The generated harmonic signals are collected by the same objective, collimated by a collimating lens into a parallel beam, and then spectrally separated by the dichroic mirror to acquire fluorescence signals in two separate channels. These signals are subsequently directed by the condenser lens onto the detection surface of a photomultiplier tube (PMT).

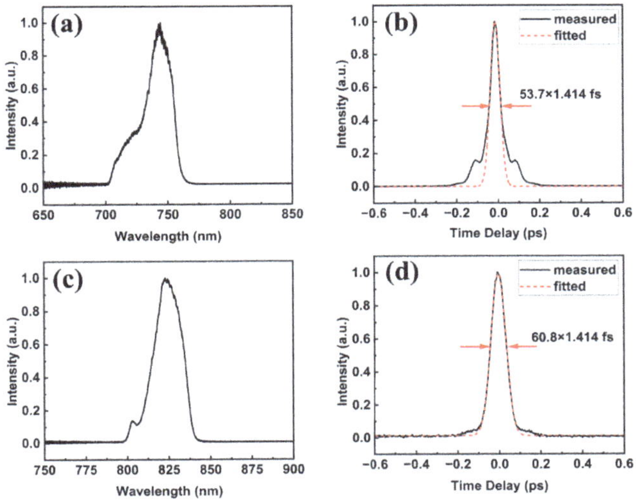

Figure 5. Output characteristics of the filtered 740 nm and 820 nm pulses. (**a**) Optical spectrum of the 740 nm pulse. (**b**) Autocorrelation trace of the 740 nm femtosecond pulse (black solid line) and the Gaussian fit (red dashed line). (**c**) Optical spectrum of the 820 nm pulse. (**d**) Autocorrelation trace of the 820 nm femtosecond pulse (black solid line) and the Gaussian fit (red dashed line).

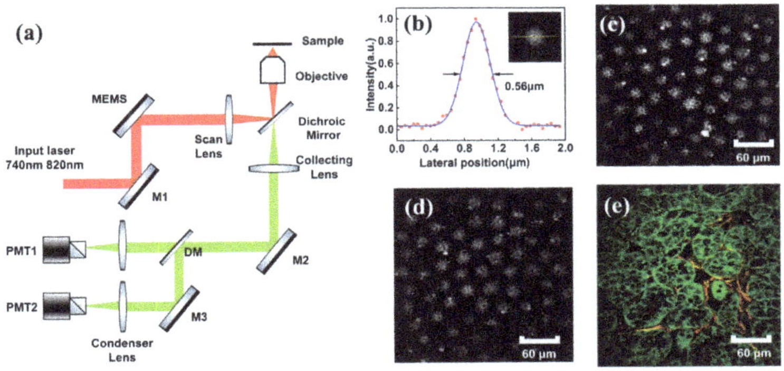

Figure 6. Schematic of the two-photon microscopy and achieved imaging results. (**a**) Schematic of the custom-built inverted two-photon microscope. (**b**) System resolution test using 100 nm fluorescent beads under 740 nm excitation, indicating a lateral resolution of 0.56 μm. (**c**) Two-photon fluorescence image of a pollen excited at 740 nm. (**d**) Two-photon fluorescence image of the pollen excited at 820 nm. (**e**) Multimodal image of a fibrotic mouse kidney section, where autofluorescence (excited at 740 nm) is shown in green, and second-harmonic generation from collagen (excited at 820 nm) is shown in red.

Figure 6c presents an image of pollen acquired using 10 mW excitation at 740 nm, clearly revealing the fine structural details. Under the same excitation power, imaging with 820 nm pulses (Figure 6d) also delineates distinct pollen features. To further evaluate the imaging performance of our dual-wavelength femtosecond laser, we performed two-photon imaging of mouse fibrotic kidney sections. Firstly, autofluorescence from the tissue was excited using the 740 nm output, and the signal was collected through a 390 nm long-pass filter, revealing key renal structures in a green pseudo color (Figure 6e). The resulting images clearly resolved the glomeruli and surrounding renal tubules. Subsequently, the sample was imaged using the 820 nm output to excite SHG from collagen fibers. The

SHG signal was isolated with a 410 nm bandpass filter, visualizing the collagen distribution in a red pseudo color (Figure 6e). Thus, the 740 nm channel was used to observe the glomerular architecture, while the 820 nm channel targeted collagen in the fibrotic kidney, demonstrating the system's capability for multimodal imaging.

4. Discussion

The demonstrated quartz-encapsulated femtosecond fiber laser provides a compact, stable, and versatile excitation source for two-photon microscopy. By delivering femtosecond pulses at 740 nm and 820 nm, the system directly covers the optimal two-photon excitation ranges of two major endogenous fluorophores involved in cellular metabolism (NADH and FAD), thereby enabling label-free metabolic imaging. The successful imaging of pollen grains and mouse kidney sections validates not only the optical performance of the laser but also its suitability for resolving biologically relevant structures at subcellular scales. These results suggest that, in specific non-laboratory application scenarios, such as portable [24] or space-constrained imaging environments [25,26], this quartz-encapsulated femtosecond fiber laser offers distinct advantages over conventional Ti:sapphire systems, including compactness, robustness, and reduced maintenance requirements.

In addition to the proof-of-concept demonstration, our study highlights several potential application scenarios for two-photon biological imaging. The dual-wavelength design is particularly well suited for optical redox ratio measurements, which are increasingly recognized as a powerful tool for assessing cellular metabolic states and detecting early pathological changes, such as cancer progression or tissue hypoxia [27–29]. By enabling selective excitation of NADH and FAD at their respective optimal wavelengths, the system enhances the accuracy of metabolic imaging compared with single-wavelength approaches. The collagen-derived second-harmonic generation signal detected at 820 nm highlights the potential of this light source for investigating extracellular matrix remodeling at both the cellular and tissue levels [30,31]. Quartz encapsulation further enhances the system by ensuring mechanical robustness, environmental stability, and long-term alignment-free operation. These features are crucial for enabling in vivo animal imaging and clinical endoscopic physiological monitoring [32].

Future research could focus on expanding the system to achieve higher pulse energies or adjustable repetition rates, which would enable multi-parameter imaging and the simultaneous multi-photon excitation of additional endogenous or exogenous fluorophores. Such enhancements would broaden the versatility of the platform, allowing more comprehensive metabolic and structural studies across diverse biological samples and experimental conditions.

5. Conclusions

In this study, we report on a solid-like, quartz-encapsulated femtosecond laser source, with a repetition rate of 80 MHz, delivering synchronous dual-wavelength pulses at 740 nm and 820 nm. The systematic long-term stability and robustness were realized by encapsulating the employed free-space optics within a sealed quartz module. Based on the home-built 80 MHz, 347 mW, 52.4 fs, 1560 nm laser pulses, the 80 MHz, 30 mW, 53.7 fs, and 740 nm laser pulses and 80 MHz, 15 mW, 60.8 fs, and 820 nm laser pulses can be delivered by this quartz-encapsulated module. The measured RMS of the output power was 0.17%@1h, indicating a perfect operating performance. We have validated the practical utility of this source by conducting two-photon imaging on mouse kidney sections and pollen samples, successfully resolving distinct biological structures through autofluorescence and second-harmonic generation. The results demonstrate that this compact and highly stable laser system is a powerful and versatile tool for advanced biophotonic applications, particularly for multimodal two-photon microscopy in metabolic studies.

Author Contributions: Conceptualization, Y.L. and A.W.; methodology, L.F.; software, B.Y.; validation, B.Y. and S.W.; formal analysis, B.Y.; investigation, S.W.; resources, A.W.; data curation, B.Y.; writing—original draft preparation, B.Y. and S.W.; writing—review and editing, Y.L. and A.W.; visualization, B.Y.; supervision, L.F.; project administration, Y.L.; funding acquisition, L.F. and Y.L. All authors have read and agreed to the published version of the manuscript.

Funding: This research was funded by the National Natural Science Foundation of China (No. 62475008, No. 62305186).

Institutional Review Board Statement: The study did not require ethical approval.

Informed Consent Statement: Not applicable.

Data Availability Statement: The data that support the findings of this study are available from the corresponding authors upon reasonable request.

Acknowledgments: We thank the Third Hospital of Sun Yat-sen University for providing fibrotic mouse kidney slices.

Conflicts of Interest: The authors declare no conflicts of interest.

Abbreviations

The following abbreviations are used in this manuscript:

SHG	Second-harmonic generation
SPM	Self-phase modulation
TPM	Two-photon microscopy
NADH	Nicotinamide adenine dinucleotide
FAD	Flavin adenine dinucleotide
PM	Polarization maintenance
PPLN	Periodically poled lithium niobate
NLAM	Nonlinear amplifying loop mirror
HNPCF	High nonlinearity photonic crystal fiber
GVD	Group velocity dispersion
FWHM	Full width at half maximum
RMS	Root-mean-square
MEMS	Micro-Electro-Mechanical System
PMT	Photomultiplier tube

References

1. Zong, W.; Wu, R.; Chen, S.; Wu, J.; Wang, H.; Zhao, Z.; Chen, G.; Tu, R.; Wu, D.; Hu, Y. Miniature Two-Photon Microscopy for Enlarged Field-of-View, Multi-Plane and Long-Term Brain Imaging. *Nat. Methods* **2021**, *18*, 46–49. [CrossRef]
2. Zhao, C.; Chen, S.; Zhang, L.; Zhang, D.; Wu, R.; Hu, Y.; Zeng, F.; Li, Y.; Wu, D.; Yu, F. Miniature Three-Photon Microscopy Maximized for Scattered Fluorescence Collection. *Nat. Methods* **2023**, *20*, 617–622. [CrossRef]
3. Bird, D.; Gu, M. Two-Photon Fluorescence Endoscopy with a Micro-Optic Scanning Head. *Opt. Lett.* **2003**, *28*, 1552–1554. [CrossRef]
4. Harzic, R.L.; Riemann, I.; Weinigel, M.; König, K.; Messerschmidt, B. Rigid and High-Numerical-Aperture Two-Photon Fluorescence Endoscope. *Appl. Opt.* **2009**, *48*, 3396–3400. [CrossRef]
5. Wang, B.G.; König, K.; Halbhuber, K.J. Two-Photon Microscopy of Deep Intravital Tissues and Its Merits in Clinical Research. *J. Microsc.* **2010**, *238*, 1–20. [CrossRef]
6. Obeidy, P.; Tong, P.L.; Weninger, W. Research Techniques Made Simple: Two-Photon Intravital Imaging of the Skin. *J. Investig. Dermatol.* **2018**, *138*, 720–725. [CrossRef]

7. Kittler, H.; Pehamberger, H.; Wolff, K.; Binder, M. Diagnostic Accuracy of Dermoscopy. *Lancet Oncol.* **2002**, *3*, 159–165. [CrossRef]
8. Jartarkar, S.R.; Patil, A.; Wollina, U.; Gold, M.H.; Stege, H.; Grabbe, S.; Goldust, M. New Diagnostic and Imaging Technologies in Dermatology. *J. Cosmet. Dermatol.* **2021**, *20*, 3782–3787. [CrossRef]
9. Diaspro, A.; Chirico, G.; Collini, M. Two-Photon Fluorescence Excitation and Related Techniques in Biological Microscopy. *Q. Rev. Biophys.* **2005**, *38*, 97–166. [CrossRef]
10. Drobizhev, M.; Molina, R.S.; Hughes, T.E. Characterizing the Two-Photon Absorption Properties of Fluorescent Molecules in the 680–1300 Nm Spectral Range. *Bio-Protocol* **2020**, *10*, e3498. [CrossRef]
11. Huang, L.; Zhou, X.; Tang, S. Optimization of Frequency-Doubled Er-Doped Fiber Laser for Miniature Multiphoton Endoscopy. *J. Biomed. Opt.* **2018**, *23*, 126503. [CrossRef] [PubMed]
12. Cao, R.; Wallrabe, H.; Periasamy, A. Multiphoton Flim Imaging of Nad (P) H and Fad with One Excitation Wavelength. *J. Biomed. Opt.* **2020**, *25*, 014510. [CrossRef] [PubMed]
13. Perry, S.W.; Burke, R.M.; Brown, E.B. Two-Photon and Second Harmonic Microscopy in Clinical and Translational Cancer Research. *Ann. Biomed. Eng.* **2012**, *40*, 277–291. [CrossRef] [PubMed]
14. Hato, T.; Winfree, S.; Day, R.; Sandoval, R.M.; Molitoris, B.A.; Yoder, M.C.; Wiggins, R.C.; Zheng, Y.; Dunn, K.W.; Dagher, P.C. Two-Photon Intravital Fluorescence Lifetime Imaging of the Kidney Reveals Cell-Type Specific Metabolic Signatures. *J. Am. Soc. Nephrol.* **2017**, *28*, 2420–2430. [CrossRef]
15. Qin, Y.; Xia, Y. Simultaneous Two-Photon Fluorescence Microscopy of Nadh and Fad Using Pixel-to-Pixel Wavelength-Switching. *Front. Phys.* **2021**, *9*, 642302. [CrossRef]
16. Fisher, W.; Wachter, E.; Armas, M.; Seaton, C. Titanium: Sapphire Laser as an Excitation Source in Two-Photon Spectroscopy. *Appl. Spectrosc.* **1997**, *51*, 218–226. [CrossRef]
17. Bindewald-Wittich, A.; Han, M.; Schmitz-Valckenberg, S.; Snyder, S.R.; Giese, G.n.; Bille, J.F.; Holz, F.G. Two-Photon–Excited Fluorescence Imaging of Human Rpe Cells with a Femtosecond Ti: Sapphire Laser. *Investig. Ophthalmol. Vis. Sci.* **2006**, *47*, 4553–4557. [CrossRef]
18. Unruh, J.R.; Price, E.S.; Molla, R.G.; Stehno-Bittel, L.; Johnson, C.K.; Hui, R. Two-Photon Microscopy with Wavelength Switchable Fiber Laser Excitation. *Opt. Express* **2006**, *14*, 9825–9831. [CrossRef]
19. Hage, C.-H.; Gomes, J.-T.; Bardet, S.; Granger, G.; Jossent, M.; Lavoute, L.; Gaponov, D.; Fevrier, S. Two-Photon Microscopy with a Frequency-Doubled Fully Fusion-Spliced Fiber Laser at 1840 Nm. *Opt. Lett.* **2018**, *43*, 5098–5101. [CrossRef]
20. Chou, L.-T.; Liu, Y.-C.; Zhong, D.-L.; Lin, W.-Z.; Hung, H.-H.; Chan, C.-J.; Chen, Z.-P.; Chia, S.-H. Low Noise, Self-Phase-Modulation-Enabled Femtosecond Fiber Sources Tunable in 740–1236 Nm for Wide Two-Photon Fluorescence Microscopy Applications. *Biomed. Opt. Express* **2021**, *12*, 2888–2901. [CrossRef]
21. Jiang, T.; Cui, Y.; Lu, P.; Li, C.; Wang, A.; Zhang, Z. All Pm Fiber Laser Mode Locked with a Compact Phase Biased Amplifier Loop Mirror. *IEEE Photonics Technol. Lett.* **2016**, *28*, 1786–1789. [CrossRef]
22. Li, M.; Yang, W.; Zhang, Z.; Wang, A. Mode-Locked Femtosecond 910 Nm Nd: Fibre Laser with Phase Biased Non-Linear Loop Mirror. *Electron. Lett.* **2017**, *53*, 1479–1481. [CrossRef]
23. Helmchen, F.; Tank, D.W.; Denk, W. Enhanced Two-Photon Excitation through Optical Fiber by Single-Mode Propagation in a Large Core. *Appl. Opt.* **2002**, *41*, 2930–2934. [CrossRef] [PubMed]
24. Wang, J.; Zhen, Z.; Wang, Y.; Wu, R.; Hu, Y.; Fu, Q.; Li, Y.; Xin, B.; Song, J.; Li, J. Non-Invasive Skin Imaging Assessment of Human Stress During Head-Down Bed Rest Using a Portable Handheld Two-Photon Microscope. *Front. Physiol.* **2022**, *13*, 899830. [CrossRef]
25. Meng, X.; Chen, J.; Zhang, Z.; Li, K.; Li, J.; Yu, Z.; Zhang, Y. Non-Invasive Optical Methods for Melanoma Diagnosis. *Photodiagnosis Photodyn. Ther.* **2021**, *34*, 102266. [CrossRef]
26. Diaspro, A.; Bianchini, P.; Vicidomini, G.; Faretta, M.; Ramoino, P.; Usai, C. Multi-Photon Excitation Microscopy. *Biomed. Eng. Online* **2006**, *5*, 36. [CrossRef]
27. Im, J.; Rajaram, N. Optical Molecular Imaging and Spectroscopy of Oxygenation and Metabolism in Tumors. *IEEE J. Sel. Top. Quantum Electron.* **2015**, *22*, 78–87. [CrossRef]
28. Georgakoudi, I.; Quinn, K.P. Optical Imaging Using Endogenous Contrast to Assess Metabolic State. *Annu. Rev. Biomed. Eng.* **2012**, *14*, 351–367. [CrossRef]
29. Kolenc, O.I.; Quinn, K.P. Evaluating Cell Metabolism through Autofluorescence Imaging of Nad (P) H and Fad. *Antioxid. Redox Signal.* **2019**, *30*, 875–889. [CrossRef]
30. Yu, B.-S.; Cheng, Q.-D.; Liu, Y.-Z.; Zhang, R.; Li, D.-W.; Wang, A.-M.; Feng, L.-S.; Jia, X. Nonlinear Imaging Detection of Organ Fibrosis in Minute Samples for Early Stage Utilizing Dual-Channel Two-Photon and Second-Harmonic Excitation. *Biosensors* **2025**, *15*, 357. [CrossRef]

31. Schenke-Layland, K. Non-Invasive Multiphoton Imaging of Extracellular Matrix Structures. *J. Biophotonics* **2008**, *1*, 451–462. [CrossRef]
32. Tanaka, K.; Toiyama, Y.; Inoue, Y.; Uchida, K.; Araki, T.; Mohri, Y.; Mizoguchi, A.; Kusunoki, M. Intravital Imaging of Gastrointestinal Diseases in Preclinical Models Using Two-Photon Laser Scanning Microscopy. *Surg. Today* **2013**, *43*, 123–129. [CrossRef]

photonics

MDPI

Article

Switchable Dual-Comb Fiber Laser Based on Sagnac Loop for Wavelength-Multiplexing and Polarization-Multiplexing

Qianyu Yao, Lilin Luo, Yue Cai, Yongguo Zheng and Xinhai Zhang *

Department of Electrical and Electronic Engineering, Southern University of Science and Technology, Shenzhen 518055, China; 12431239@mail.sustech.edu.cn (Q.Y.); 12332157@mail.sustech.edu.cn (L.L.); caiy3@sustech.edu.cn (Y.C.); 12233207@mail.sustech.edu.cn (Y.Z.)
* Correspondence: zhangxh@sustech.edu.cn

Abstract

We present an all-fiber erbium-doped mode-locked laser capable of switching among continuously tunable single-pulse mode-locking, wavelength-multiplexing asynchronous-pulse mode-locking, and polarization-multiplexing asynchronous-pulse mode-locking states. The multiplexing mechanisms under different conditions are confirmed by separating the asynchronous-pulse sequences. Experimental results and numerical simulations indicate that the adjustment of the polarization controller within the Sagnac loop is the key factor for switching between wavelength- and polarization-multiplexing asynchronous-pulse mode-locking. The multiple output characteristics of the same laser can support diverse application scenarios, offering significant cost reduction in practical applications. To the best of our knowledge, this is the first demonstration of switching between wavelength- and polarization-multiplexing asynchronous-pulse mode-locking states in a noise-like laser. Compared to previous related work, the proposed laser not only enables tunable mode-locking wavelengths but also achieves higher pulse energy. This work provides a light source solution with a simple structure and high switchability for dual-comb applications.

Keywords: switchable dual-comb fiber laser; wavelength-multiplexing asynchronous-pulse; polarization-multiplexing asynchronous-pulse

1. Introduction

Optical frequency combs manifest as periodic trains of ultrashort pulses in the time domain, and as a series of narrow-linewidth spectral lines with equal frequency spacing, fixed positions, and broad spectral coverage in the frequency domain. Highly stable optical frequency combs are extensively applied in high-precision ranging [1], optical atomic clocks [2], and precision spectroscopy [3]. Conventional dual-comb systems usually require two independent lasers with phase locking or additional modulation techniques [4–8], which greatly increases the cost and complexity of the system. In recent years, researchers have introduced the concept of "multiplexing" into the field of fiber lasers, making single-cavity dual-comb mode-locked fiber lasers a new research focus. Such lasers can generate dual-comb pulse sequence from the same cavity through wavelength multiplexing [9–13], polarization multiplexing [14–16], and propagation-direction multiplexing [17], which not only suppresses common-mode noise but also significantly simplifies the structure of dual-comb systems.

Single-cavity dual-comb mode-locked fiber lasers based on different multiplexing mechanisms possess distinct advantages. For example, in wavelength multiplexing, light waves at different wavelengths propagate with different group velocities in the fiber, usually leading to a larger repetition rate difference, which is beneficial for achieving higher sampling rates in asynchronous optical sampling [18,19]. Meanwhile, the generated asynchronous pulse sequences can be conveniently separated using wavelength-division multiplexers (WDMs). In polarization multiplexing, the two pulse sequences have a smaller difference in central wavelength and exhibit higher spectral overlap, thereby avoiding the need for spectral broadening in dual-comb ranging [15]. However, under this multiplexing mechanism, the output pulse sequences are difficult to completely separate by polarization beam splitter (PBS). In directional multiplexing, the two pulse sequences propagate clockwise and counterclockwise in the laser cavity, allowing for automatic separation in laser outputs of different directions. Nevertheless, bidirectional propagation enhances gain competition within the cavity, resulting in reduced output stability.

In summary, if dual-pulse sequence outputs with different multiplexing schemes can be realized within the same laser cavity, they will provide more practical dual-comb light sources for diverse applications. In 2016, Wu et al. [20] first reported the coexistence of scalar and vector solitons in a fiber laser mode-locked based on the nonlinear polarization rotation (NPR) mechanism, demonstrating that two different types of nonlinear pulses can exist within the same laser cavity. In the same year, Wu et al. [21] achieved switchable generation of vector and scalar solitons in a fiber laser cavity by utilizing a saturable absorber and the residual polarization dependent loss (PDL) in a wavelength-division multiplexer (WDM). These works laid the theoretical foundation for subsequent generation of different types of dual-pulse sequences within a single laser cavity. In 2019, Li et al. [22] achieved both wavelength- and polarization-multiplexing asynchronous pulse mode-locking by tuning the erbium-doped fiber (EDF) gain peak in combination with intracavity birefringence. However, due to the fixed EDF gain peaks near 1530 nm and 1560 nm, wavelength tuning could not be achieved. In 2024, Jin et al. [23] demonstrated switchable wavelength- and polarization-multiplexing asynchronous pulse mode-locking in an EDF laser. However, the resulting asynchronous pulses exhibited relatively low pulse energies. Wavelength-multiplexing dual-comb generation mechanism relies on wavelength-selective devices, which often exhibit PDL that may interfere with polarization-multiplexing. The key for polarization-multiplexing is to maintain simultaneous oscillation of light in two orthogonal polarization directions. If there is PDL in the cavity, it may suppress the light in one polarization direction, thereby preventing the realization of polarization-multiplexing [23]. Therefore, coexistence of wavelength- and polarization-multiplexing within a single laser cavity remains highly challenging.

In this work, we present an erbium-doped mode-locked laser based on a Sagnac loop, which achieves tunable single-pulse mode-locking as well as wavelength- and polarization-multiplexing asynchronous pulse mode-locking through precise control of pump power and polarization controllers (PCs). To the best of our knowledge, this is the first demonstration of switching between wavelength-multiplexing and polarization-multiplexing asynchronous pulse mode-locking in a noise-like laser (NLL). The filtering transmission spectrum of the Sagnac loop exhibits a comb-like distribution and is tunable, which has led to its widespread application in dual-wavelength and multi-wavelength mode-locked fiber lasers in previous studies [9,24]. Experimental results demonstrate that a Sagnac loop-based fiber laser, through the polarization-maintaining fiber (PMF) within the loop, can achieve polarization-multiplexing asynchronous pulse mode-locking. Compared to previous related works, the proposed laser not only offers the advantage of wavelength tunability but also delivers higher pulse energy due to the generation of noise-like pulses (NLPs). Overall, this work provides a potential light source with a simple architecture and strong switching capability for dual-comb systems.

2. Experimental Setup and Principle

The architecture of the laser constructed in this work is shown in Figure 1a. The pump source is a laser diode operating at a center wavelength of 976 nm and a maximum output power of 400 mW. The pump source and the reflection-type packaged SESAM (BATOP, SAM-1550-17-1.5ps, Jena, Germany) are coupled into the cavity through a 980/1550 nm WDM, which is connected to the Sagnac loop via a 60 cm-long EDF (LIEKKI, Er110-4/125, Camas, WA, USA). A polarization controller (PC1) is used in the optical path to adjust the polarization state of the light wave before it enters the Sagnac loop. The Sagnac loop consists of a 50/50 optical coupler (OC1), a 45 cm-long PMF (Corning, PM 1550, Corning, NY, USA) with birefringence of ~3.9×10^{-4}, another polarization controller (PC2), and a 10/90 optical coupler (OC2), where the 10/90 OC2 is responsible for outputting 10% of the cavity energy.

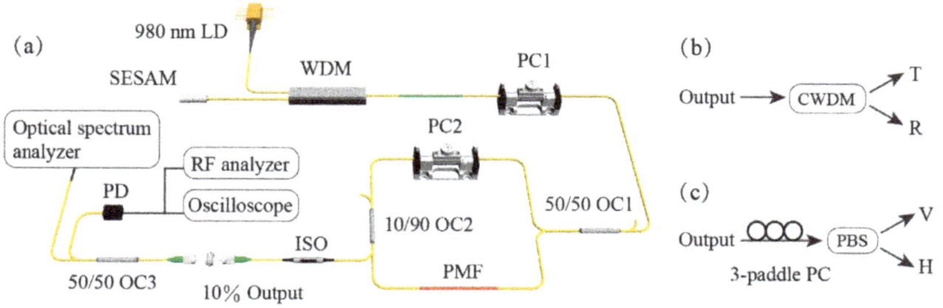

Figure 1. (**a**) Schematic diagram of the laser architecture. (**b**) Schematic diagram of asynchronous pulse sequence separation based on wavelength multiplexing and (**c**) polarization multiplexing.

The EDF in the laser cavity has a dispersion parameter of ~−20 ps/nm/km at 1550 nm and an absorption coefficient of ~110 dB/m at 1530 nm. The dispersion parameters of the PMF and the single-mode fiber (SMF) at 1550 nm are ~16 ps/nm/km and ~18 ps/nm/km, respectively. The total cavity length of the laser is 12.65 m, and the calculated net cavity dispersion is approximately -0.23 ps^2, indicating that the laser operates in the negative dispersion region. The output pulses of the laser were measured using an oscilloscope (RIGOL, DS6104, Suzhou, China) combined with a photodetector (Thorlabs DET08CFC/M, Newton, MA, USA). The optical spectrum and radio-frequency (RF) spectrum were recorded using an optical spectrum analyzer (YOKOGAWA, AQ6370B, Tokyo, Japan) and an RF analyzer (RIGOL, DSA815, Suzhou, China), respectively. The output pulse width was measured using an autocorrelator (FEMTOCHROME, FR-103XL, Berkeley, CA, USA).

When the light wave enters the Sagnac loop through the OC1, part of the light propagates clockwise while the other part propagates counterclockwise. In the propagation, the two beams accumulate a phase difference and recombine at the OC1, where interference occurs, thereby producing a filtering effect. When considering only the filtering effect of the Sagnac loop, its transmittance curve can be expressed as [25]:

$$T = \left(\sin\theta \cos\left(\frac{\phi}{2}\right)\right)^2, \tag{1}$$

where θ represents the polarization rotation angle controlled by the PC2, and $\phi = 2\pi BL/\lambda$ represents the phase difference introduced by the PMF. Here, B is the birefringence of the PMF, and L is the length of the PMF. The free spectral range (FSR) of the transmission curve of the Sagnac loop is given by [9]:

$$\Delta\lambda = \lambda^2/BL, \tag{2}$$

when the PMF length L in the optical path is 45 cm, the calculated FSR of the transmission curve of the Sagnac loop at 1570 nm is 14.05 nm. According to this formula with a fixed B value, different PMF lengths produce different FSRs of the transmission curve.

3. Results and Discussion

3.1. Single Pulse Mode-Locking with Tunable Center Wavelength

When the pump power is set to 85 mW, the laser achieves stable single pulse mode-locking, with an output power of 0.63 mW. On this basis, by adjusting the PC2 within the Sagnac loop, the center wavelength of the single pulse sequence output can be continuously tuned from 1565.9 nm to 1578.1 nm. Adjusting the PC2 changes the polarization state of the light propagating in the fiber, thus introducing additional phase shift between the two counter-propagating light beams inside the Sagnac loop. As the two beams pass through the OC1 again, the interference condition changes accordingly, leading to a shift in the peak position of the transmission curve of the Sagnac loop. This mechanism enables continuous tuning of the center wavelength of the single pulse sequence output within a certain wavelength range.

Figure 2 shows the output characteristics of single-pulse sequence at three representative wavelengths. As shown in Figure 2a, the center wavelengths of the typical optical spectra are at λ_1 = 1565.9 nm, λ_2 = 1570.0 nm, and λ_3 = 1576.8 nm, with −3 dB bandwidths of 2.07 nm, 2.14 nm, and 1.70 nm, respectively. The −3 dB bandwidth represents the frequency range where the signal power is reduced to half of its maximum value, and a signal within this range is considered to have sufficient energy and intensity. Figure 2b presents the corresponding RF spectra of the typical optical spectra, where the repetition rates corresponding to λ_1, λ_2, and λ_3 are 15.809571 MHz, 15.809467 MHz, and 15.809306 MHz, respectively. The length of the laser cavity in this work is 12.65 m. Based on the formula $f = c/nL$, it can be determined that the laser is operating in the mode-locking state of fundamental frequency. From Figure 2b, it can be observed that the repetition rate of the mode-locked pulses decreases as the center wavelength increases. This phenomenon is attributed to the laser operation in the negative dispersion regime, where the group velocity of the light with longer-wavelength is lower than that of the light with shorter-wavelength. Figure 2c,d show the corresponding pulse train and the autocorrelation trace when the center wavelength of the mode-locked pulses is at λ_2. From Figure 2c, it can be seen that the pulse train has a period of 63 ns, corresponding to the time of a pulse propagating in the laser cavity for a round trip, indicating that it is mode-locked at the fundamental mode-locking frequency. The autocorrelation trace of the output pulses shown in Figure 2d features a narrow spike sitting on a picosecond-scale pedestal. The double-scale autocorrelation trace together with the smooth spectrum indicates that the output corresponds to NLPs. The width of the pedestal characterizes the duration of the NLP wave packet. In this state, as the pump power increases, the top of the pedestal gradually flattens, which can be attributed to the further broadening of the NLP wave packet such that more portions extend beyond the scanning range of the autocorrelator.

We further investigated the stability of the laser in the single-pulse mode-locked state. Figure 3 shows the variations in the laser's spectrum, output power, and repetition rate during 12 h of continuous operation in this mode-locked state. As shown in Figure 3, the output spectrum and power of the laser remained essentially unchanged during the 12-h operation, demonstrating good stability. The maximum drift of the repetition rate during this period was 66 Hz, which we speculate is primarily caused by changes in the environmental temperature.

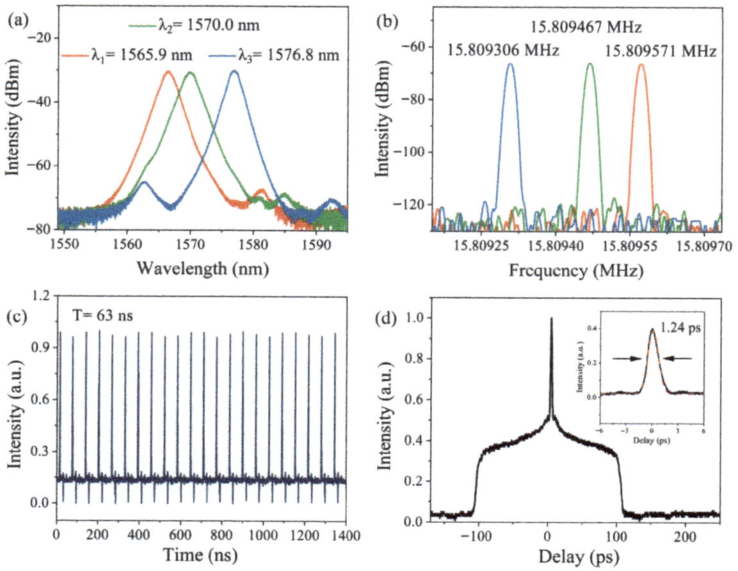

Figure 2. Output characteristics of the single-pulse sequence at three representative wavelengths: (**a**) optical spectra and (**b**) RF spectra. At the mode-locking center wavelength of λ_2: (**c**) pulse train and (**d**) autocorrelation trace.

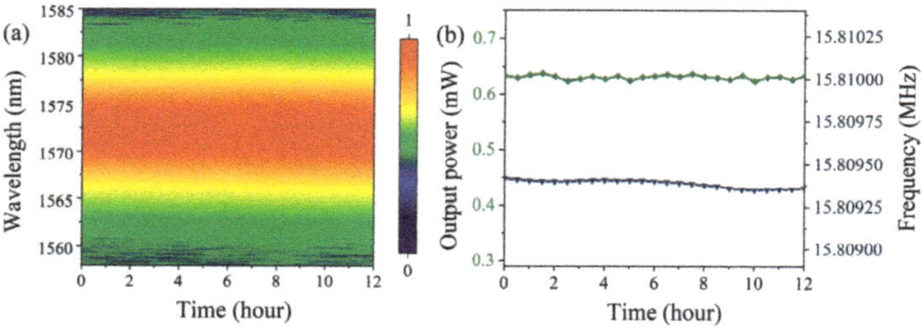

Figure 3. (**a**) Spectrum, (**b**) output power, and repetition rate during 12 h of continuous operation in the single-pulse mode-locked state.

3.2. Asynchronous Pulse Sequence Generation with Wavelength-Multiplexing

When the pump power is at 120 mW, the laser can achieve mode-locking with asynchronous pulse sequence generation, by carefully adjusting the PCs, and the mode-locking state can be maintained until the pump power is increased to 250 mW. Figure 4 shows output characteristics recorded with the pump power at 160 mW, where the output power is 1.40 mW. The pulse interference pattern with a period of 3.1 ms in Figure 4a indicates that the laser generates asynchronous pulse sequences. Setting the time window of the oscilloscope to 700 ns, two sets of pulse sequences with different amplitudes can be clearly observed. As shown in Figure 4b,c, the time intervals between the two sets of pulses vary at different moments, further demonstrating the asynchronous nature of the two pulse trains. Figure 4d,e show the RF spectra and optical spectra of the asynchronous pulse sequences, respectively. As shown in Figure 4d, the RF spectrum exhibits a distinct double-peak feature, with the peaks located at 15.809605 MHz and 15.809922 MHz, respectively, indicating a repetition rate difference of 317 Hz between the two pulse sequences. This frequency difference corresponds to the reciprocal of the pulse interference period shown in Figure 4a, with a slight deviation mainly due to the limited resolution of the RF analyzer.

The optical spectrum shown in Figure 4e has two distinct center wavelengths, located at 1563.3 nm and 1577.3 nm, with −3 dB bandwidths of 2.3 nm and 1.5 nm, respectively. The wavelength separation between the two mode-locking centers is 14 nm, corresponding to the FSR of the transmission spectrum of the Sagnac loop. From the experimental observation presented above, it can be inferred that the mechanism of the asynchronous pulse sequence generation is based on wavelength-multiplexing with the Sagnac loop filtering effect. From the relationship between the center wavelength of the pulse sequence and repetition rate discussed in Section 3.1, it can be concluded that the pulse sequence with the center wavelength at λ_1 has the repetition rate of 15.809922 MHz, while the pulse sequence with the center wavelength of λ_2 has the repetition rate of 15.809605 MHz. Finally, the autocorrelation trace shown in Figure 4f indicates that the pulse is noise-like, and fitting with Gaussian yields a spike width of 1.09 ps.

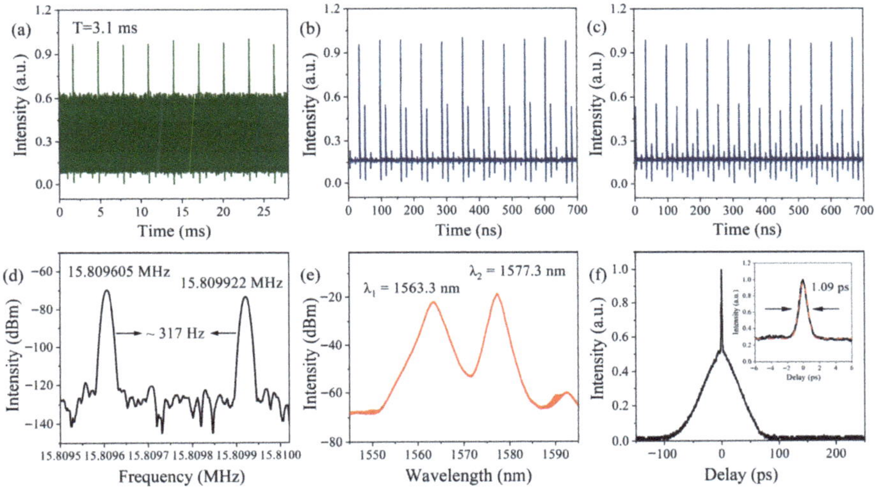

Figure 4. Output characteristics of the wavelength-multiplexed asynchronous pulse sequences: (**a**) pulse interference pattern; (**b**,**c**) pulse trains at different moments; (**d**) RF spectrum; (**e**) optical spectrum; and (**f**) autocorrelation trace.

We measured the stability of the laser in the wavelength-multiplexing asynchronous mode-locked state. Figure 5 shows the variations in the spectrum, output power, repetition rate, and repetition rate difference of the laser during 12 h of continuous operation in this mode-locked state. As shown in Figure 5a,b, the output spectrum and power exhibited no significant changes during the 12-h operation, demonstrating good stability. As illustrated in Figure 5c, the repetition rates of the two asynchronous pulse sequences each experienced a maximum drift of about 60 Hz; however, in the absence of any external modulation devices, the maximum drift of their repetition rate difference remained within 4 Hz. This result indicates that the inherent common-cavity effect of the single-cavity dual-comb system can effectively suppress the influence of environmental perturbations on the repetition rate difference between the two combs, thus providing important assurance for the stability and reliability of dual-comb light sources.

To further confirm that the asynchronous pulse sequences were generated via wavelength-multiplexing, we connected a coarse wavelength-division multiplexer (CWDM) at the laser output, as shown in Figure 1b. The CWDM has three ports: the COM input port, the transmission output port (1539–1565 nm), and the reflection output port (1575–1620 nm).

After separation by the CWDM, the powers at the two output ports are 0.75 mW and 0.49 mW, respectively.

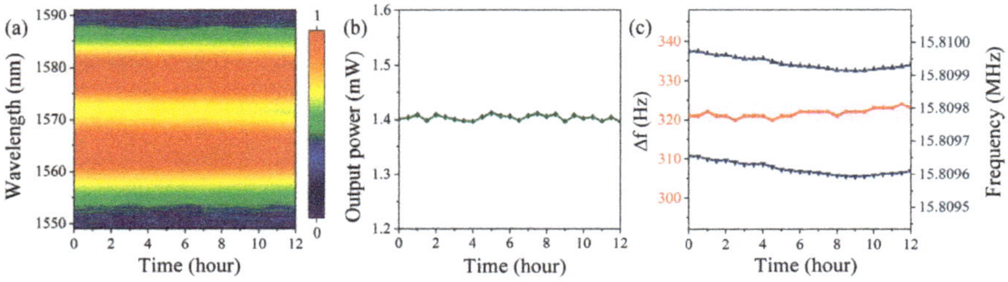

Figure 5. (**a**) Spectrum, (**b**) output power, and (**c**) repetition rate and repetition rate difference during 12 h of continuous operation in the wavelength-multiplexing asynchronous mode-locked state.

As shown in Figure 6a, the green curve represents the optical spectrum measured from the transmission port of the CWDM while the blue curve shows the optical spectrum measured from the reflection port of the CWDM. After passing through the CWDM, the two pulse sequences are basically separated. The slight overlap of the two spectra is due to the insufficient isolation between the two output ports of the CWDM. Figure 6b shows the RF spectra measured from the two output ports of the CWDM. From the RF spectrum, it can be seen that the pulse sequence with the center wavelength of λ_2 is completely filtered out at the transmission output port, while a portion of the pulse sequence with the center wavelength of λ_1 still remains at the reflection output port. At the reflection port, the signal-to-noise ratio (SNR) of the pulse sequence with the center wavelength of λ_2 is significantly higher than that of λ_1, with an SNR difference of more than 30 dB. Therefore, the pulse sequence with the higher repetition rate from the reflection output can be regarded as noise. Figures 6c and 6d show the pulse sequences measured from the transmission and reflection output ports of the CWDM, respectively. As illustrated, the asynchronous pulse sequences generated in the laser cavity can be successfully separated with the CWDM, further confirming that the generation of the asynchronous pulse sequences is wavelength-multiplexing based. Specifically, this is accomplished by leveraging the interference effect of the Sagnac loop to create varying transmissivity for light waves of different wavelengths, thereby forming a filtering effect and realizing wavelength-multiplexing generation of the asynchronous pulse sequences.

3.3. Asynchronous Pulse Sequence Generation with Polarization-Multiplexing

When the pump power is set to 190 mW, a pulse interference pattern with a larger fringe spacing than before is observed during the adjustment of the PCs. At this point, the output power of the laser is 1.72 mW. As shown in Figure 7a, the pulse interference period now is 8.4 ms. Together with the two sets of pulse sequences observed at different moments shown in Figure 7b,c, the results indicate that the laser now generates asynchronous pulse sequences with a different repetition rate difference than before. From the RF spectrum in Figure 7d, the repetition rates of the two pulse sequences in this state are determined to be 15.809628 MHz and 15.809801 MHz, respectively, corresponding to a repetition rate difference of 119 Hz, which is significantly smaller than the previous value of 317 Hz. Figure 7e displays the optical spectrum of the laser output with the center wavelength of 1574.1 nm and the −3 dB bandwidth of 5.3 nm. The optical spectrum exhibits a pronounced feature of the asymmetric sidebands, a phenomenon commonly observed in single-wavelength mode-locked pulse sequences in previous studies [26–29]. Our experiments demonstrate that the phenomenon of asymmetric spectral sidebands can also appear in asynchronous pulse sequences. Based on the spectral characteristics, it is inferred that the generation of asynchronous pulse sequences has transitioned from wavelength-multiplexing to other mechanisms. The autocorrelation trace in Figure 7f indicates that the pulse sequence in this state still consists of NLPs.

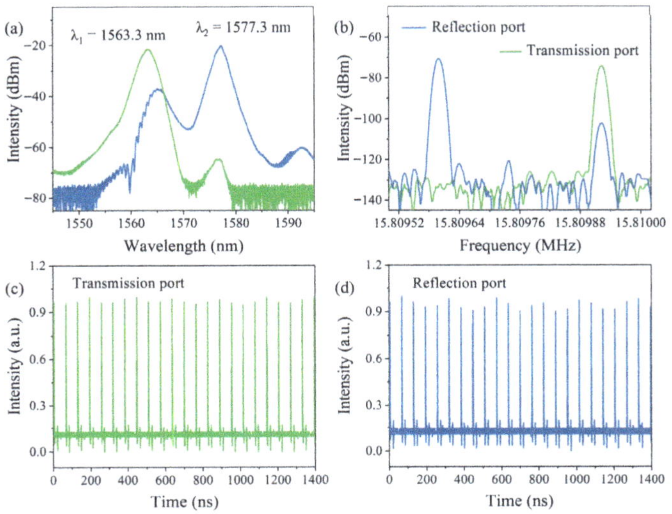

Figure 6. Pulse characteristics measured at the two output ports of the CWDM: (**a**) optical spectra, (**b**) RF spectra, and (**c**,**d**) pulse sequences.

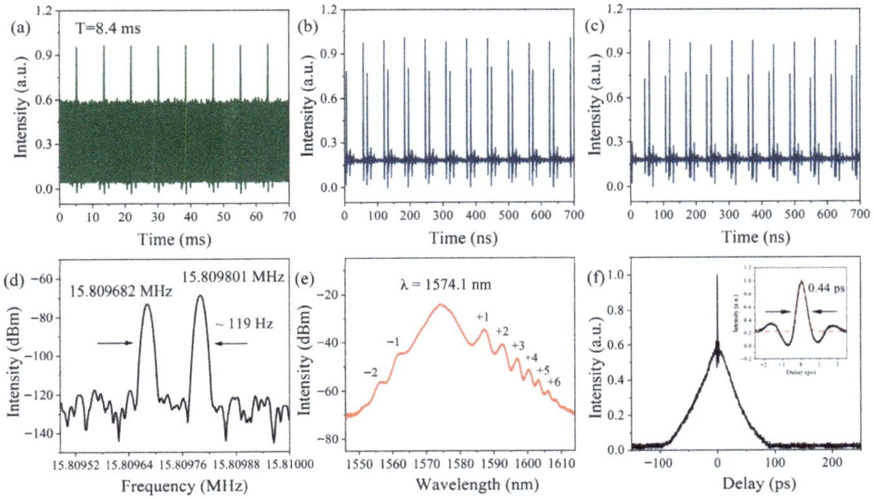

Figure 7. Output characteristics of the polarization-multiplexed asynchronous pulse sequences: (**a**) pulse interference pattern; (**b**,**c**) pulse trains at different moments; (**d**) RF spectrum; (**e**) optical spectrum; and (**f**) autocorrelation trace.

We measured the stability of the laser in the polarization-multiplexing asynchronous mode-locked state. Figure 8 shows the variations in the spectrum, output power, and repetition rate of the laser during 12 h of continuous operation in this mode-locked state. As shown in Figure 8a,b, the output spectrum and power exhibited no significant changes during the 12-h operation, demonstrating good stability. As illustrated in Figure 8c, the repetition rates of the two asynchronous pulse sequences each experienced a maximum drift of about 74 Hz; however, the maximum drift of their repetition rate difference remained within 4 Hz, indicating that the single-cavity dual-comb system can effectively suppress the influence of environmental perturbations on the repetition rate difference between the two combs.

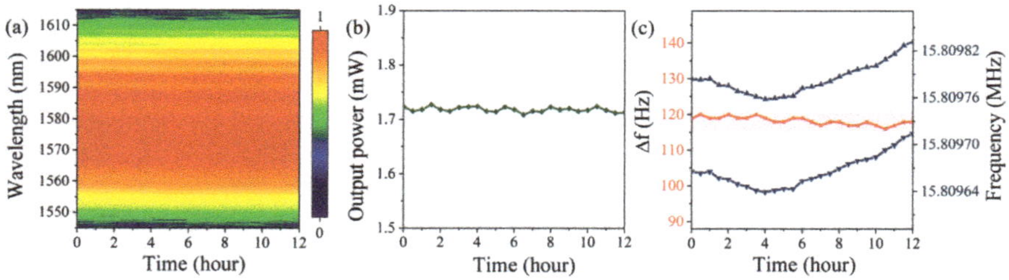

Figure 8. (**a**) Spectrum, (**b**) output power, and (**c**) repetition rate and repetition rate difference during 12 h of continuous operation in the polarization-multiplexing asynchronous mode-locked state.

To elucidate the generation mechanism of the asynchronous pulse sequences, as shown in Figure 1c, a three-paddle polarization controller (3-paddle PC) and a PBS were connected to the laser output. By adjusting the 3-paddle PC placed before the PBS, the RF signal at the low repetition rate from the horizontal output port of the PBS was maximized. At this point, the optical powers at the vertical and horizontal output ports of the PBS were measured to be 0.76 mW and 0.69 mW, respectively.

The RF spectrum after PBS separation is shown in Figure 9a. In Figure 9a, the polarization extinction ratio of the two pulses is approximately 23 dB, indicating that the two pulse sequences are nearly orthogonally polarized to each other. This confirms that the generation of the asynchronous pulse sequences in the laser now is due to polarization-multiplexing [14,16]. Figure 9b shows the optical spectra of the pulse sequences after the PBS. As illustrated, the spectra measured from the vertical and horizontal polarization ports of the PBS exhibit similar shapes, with center wavelengths of 1573.9 nm and 1574.6 nm, respectively. Figure 9c,d show the pulse sequences measured with oscilloscope from the vertical and horizontal polarization ports of the PBS, respectively. Different from the pulse sequences separated by the CWDM with wavelength-multiplexing, a small amount of the other set of pulse sequences is still visible in the separated pulse sequences. This is because the two pulse sequences are not completely orthogonally polarized. The birefringence of the PMF inside the laser cavity leads to different refractive indices along different polarization directions. Combined with the meticulous adjustment of the PCs, this induces two sets of pulse sequences with distinct repetition rates in the orthogonal directions, which fundamentally explains the phenomenon of asynchronous pulse sequence output at this time.

3.4. Switching Between Wavelength-Multiplexing and Polarization-Multiplexing

3.4.1. Experimental Analysis

To investigate the key factor for switching between the two multiplexing mechanisms, we first studied the effect of pump power on the laser in the two mode-locked states. As shown in Figure 10a, in the wavelength-multiplexing mode-locked state, changing only the pump power from 120 to 200 mW led to an increase in spectral intensity, while the spectral shape remained essentially unchanged and the asynchronous mode-locking was maintained throughout. Similarly, in the polarization-multiplexing mode-locked state shown in Figure 10b, changing only the pump power from 130 to 250 mW also increased the spectral intensity without significantly altering the spectral shape, and the asynchronous mode-locking was preserved. These results indicate that variations in pump power are not the key factor for switching between the two multiplexing mechanisms.

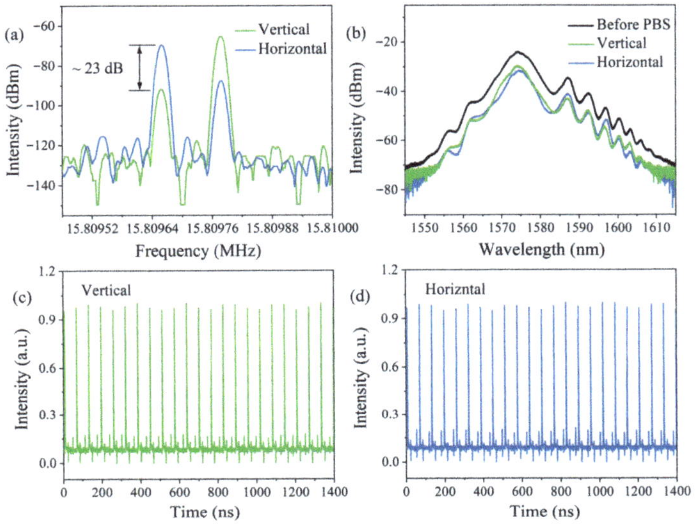

Figure 9. Pulse characteristics measured at the two output ports of the PBS: (**a**) Optical spectra, (**b**) RF spectra, and (**c**,**d**) pulse sequences.

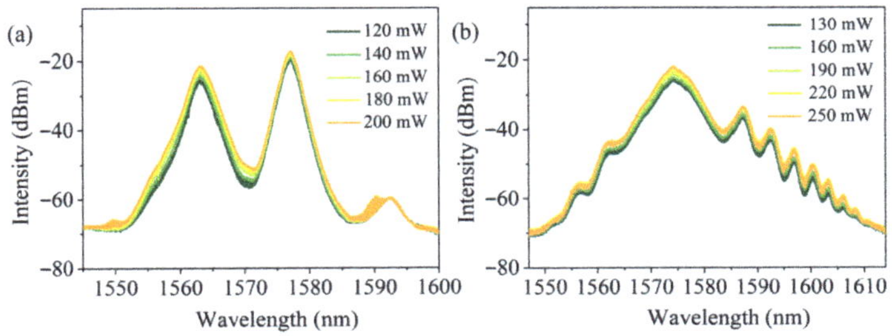

Figure 10. Optical spectral variations with pump power during (**a**) wavelength-multiplexing and (**b**) polarization-multiplexing mode-locked states.

When the pump power is set to 190 mW, the red curve in Figure 11 represents the optical spectrum under the polarization-multiplexing asynchronous mode-locking state. However, the optical spectrum changes as PC2 is adjusted. In the optical spectral diagram, the mode-locking center and the secondary sidebands begin to exhibit varying degrees of blue shift. The sidebands on both sides of the optical spectrum gradually become smooth, and the intensity of the -1 sideband to the left of the mode-locking center increases continuously as the optical spectrum blue-shifts. Eventually, as the state of PC2 changes, the mode-locking center and the -1 sideband in the original red spectral curve dominate the gain competition, forming a double-peak spectrum. Meanwhile, the laser also transitions from the polarization-multiplexing asynchronous mode-locking state to the wavelength-multiplexing asynchronous mode-locking state. From the above experimental process, it can be concluded that the adjustment of PC2 in the laser is the key factor for switching between wavelength-multiplexing and polarization-multiplexing asynchronous pulses.

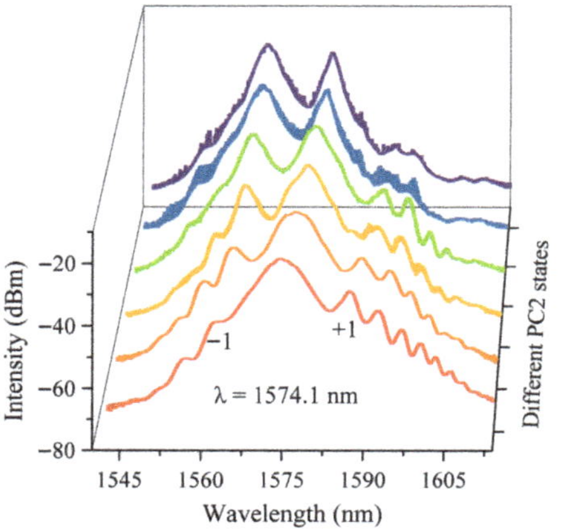

Figure 11. Optical spectral variation with PC2 adjustment at a pump power of 190 mW.

3.4.2. Numerical Analysis

To further analyze the mechanism underlying the switching between two multiplexing modes in the constructed laser, we perform a numerical simulation of the transmission curve of the Sagnac loop. Based on Equation (1), $T = \left(\sin\theta\cos\left(\frac{\phi}{2}\right)\right)^2 = \left(\sin\theta\cos\left(\frac{\pi BL}{\lambda}\right)\right)^2$, presented in Section 2, which considers only the filtering effect of the Sagnac loop, it can be concluded that the maximum transmission is achieved when $\theta = \frac{\pi}{2} + k\pi$ (k is an integer). At this condition, the FSR of the Sagnac loop transmission curve is jointly determined by the length of the PMF within the loop and its birefringence. As shown in Figure 12a,b, the FSR of the transmission curve decreases with increasing PMF length or birefringence.

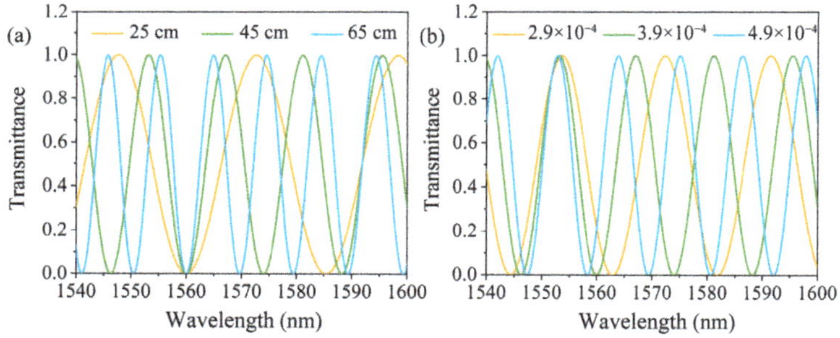

Figure 12. (**a**) Transmission spectra for different values of L at $\theta = \pi/2$, $B = 3.9 \times 10^{-4}$, (**b**) Transmission spectra for different values of B at $\theta = \pi/2$, L = 45 cm.

Based on Equation (1), further considering the additional phase delay introduced by adjusting PC2, the transmission curve of the Sagnac loop is thereby refined as [30]

$$T = \left(\sin\theta\cos\left(\frac{\phi + \phi_{PC}}{2}\right)\right)^2 \quad (3)$$

Here ϕ_{PC} represents the additional phase delay introduced by PC2. As shown in Figure 13, the additional phase delay caused by PC2 shifts the transmission curve, thereby altering the position of the transmission peaks. This is the primary reason for the experimentally observed blue shift in the spectrum when adjusting PC2. In practical operation, adjusting PC2 changes the polarization state distribution within the laser cavity. This process not only introduces an extra phase delay that modifies the transmission of the

Sagnac loop but also affects whether stable pulses can form along different polarization directions inside the cavity.

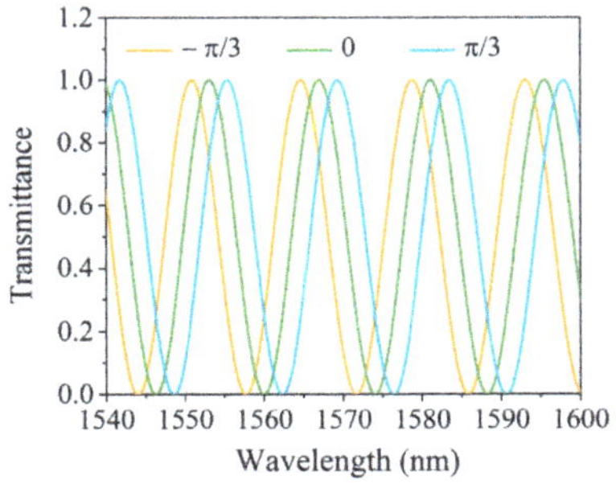

Figure 13. Transmission spectra for different values of ϕ_{PC} at $\theta = \pi/2$, $B = 3.9 \times 10^{-4}$, $L = 45$ cm.

3.4.3. Adjustment of PC2 as the Key Factor for Switching

From the above experiments and numerical analysis, it can be seen that the red curve in Figure 11 corresponds to the spectrum in the polarization-multiplexing asynchronous mode-locked state, where the transmission peak of the Sagnac loop is located near the central mode-locking wavelength of $\lambda = 1574.1$ nm. Based on the birefringence of the PMF within the Sagnac loop, precise adjustment of the pump as well as PC1 and PC2 allows light to simultaneously oscillate along two orthogonal polarization directions in the cavity, ultimately forming polarization-multiplexing asynchronous pulse mode-locking. On this basis, adjusting PC2 alters the distribution of polarization states inside the laser cavity, causing the light to no longer satisfy the condition for simultaneous oscillation along the two orthogonal polarization directions. Concurrently, due to changes in phase and transmission, a dual-peak spectrum gradually forms, and the laser transitions from polarization-multiplexing asynchronous pulse mode-locking to wavelength-multiplexing asynchronous pulse mode-locking. The purple curve in Figure 11 represents the optical spectrum under the wavelength-multiplexing asynchronous mode-locking state. However, this asynchronous mode-locking state is unstable at this point. Through precise control of the pump power and PCs, a stable wavelength-multiplexing asynchronous mode-locking state, as shown in Figure 4, can ultimately be achieved. In summary, the adjustment of PC2 within the laser is a key factor for switching between wavelength- and polarization-multiplexing asynchronous pulse sequences, but obtaining stable asynchronous pulse mode-locking still requires a combination of precise control of the pump and PC1.

4. Comparison with Previous Related Work

As shown in Table 1, a total of two lasers similar to the present work have been identified, both capable of switching between wavelength- and polarization-multiplexing asynchronous pulse mode-locking. Considering the maximum output power of the LD, the maximum pump power is set to 380 mW under single-pulse mode-locking, resulting in an output power of 3.96 mW at this point. For the stability of asynchronous pulse mode-locking, the maximum pump power is set to 250 mW during wavelength-multiplexing mode-locking, yielding an output power of 2.42 mW. After separation by the CWDM, the powers of the two pulse sequences are 1.37 mW and 0.73 mW, respectively. In the polarization-multiplexing mode-locking state, the maximum pump power is set to 270 mW, corresponding to an output power of 2.67 mW. After separation by the PBS, the powers of the two pulse sequences are 1.2 mW and 1.08 mW, respectively. In this work, a Sagnac-loop-based filtering structure was employed, allowing continuous tunability of the mode-locking wavelength by adjusting PC2. Furthermore, since the laser output in this setup consists of NLPs, the laser exhibits higher pulse energy compared to previous similar works.

Table 1. Comparison with previous related work.

Ref.	Continuous Wavelength Tunability	Single-Pulse Mode-Locking Output Pulse Energy (pJ)	Wavelength Multiplexing Output Pulse Energy (pJ)		Polarization Multiplexing Output Pulse Energy (pJ)	
			Before Pulse Separation	After Pulse Separation	Before Pulse Separation	After Pulse Separation
[22]	No	12.17	-	-	-	5.67&9.53
		10.22	-	-	-	4.03&6.27
[23]	Yes	23.18	25.75	-	48.58	-
		-	32.19	-	64.81	-
This work	Yes	39.87	88.61	31.01&47.47	108.86	43.67&48.10
		250.63	153.16	46.20&86.71	168.99	68.35&75.95

5. Conclusions

In this work, we demonstrate an erbium-doped mode-locked fiber laser based on a Sagnac loop. By precise control of the pump power and PCs, the laser can achieve tunable single-pulse mode-locking, as well as wavelength- and polarization-multiplexing asynchronous pulse mode-locking. The tunable single-pulse mode-locking and wavelength-multiplexing asynchronous pulse mode-locking primarily rely on the filtering effect of the intracavity Sagnac loop, while the polarization-multiplexing asynchronous pulse mode-locking mainly originates from the birefringence of the PMF inside the Sagnac loop. The two asynchronous pulse sequences are separated using a CWDM and a PBS, respectively, further confirming their multiplexing mechanisms. Experimental and numerical simulations are conducted to investigate the key factors governing the switching between wavelength- and polarization-multiplexing asynchronous pulses. To the best of our knowledge, this is the first demonstration of switching between wavelength- and polarization-multiplexing asynchronous pulse mode-locking in NLPs. Compared to previous related works, employing the Sagnac loop as a filter enables continuous tunability of the mode-locking wavelength, while the NLP output delivers higher pulse energy than that of conventional solitons. This work provides a potential light source with a simple architecture and strong switching capability for dual-comb systems.

Author Contributions: Writing—original draft preparation, Q.Y.; investigation, L.L., Y.C. and Y.Z.; supervision, X.Z. All authors have read and agreed to the published version of the manuscript.

Funding: This research was funded by China National Research Program (Grant 2022YFB3806900).

Institutional Review Board Statement: Not applicable.

Informed Consent Statement: Not applicable.

Data Availability Statement: The raw data supporting the conclusions of this article will be made available by the authors on request.

Conflicts of Interest: The authors declare no conflicts of interest.

References

1. Minoshima, K.; Matsumoto, H. High-accuracy measurement of 240-m distance in an optical tunnel by use of a compact femtosecond laser. *Appl. Opt.* **2000**, *39*, 5512–5517. [CrossRef]
2. Nakamura, T.; Ito, I.; Kobayashi, Y. Offset-free broadband Yb:fiber optical frequency comb for optical clocks. *Opt. Express* **2015**, *23*, 19376–19381. [CrossRef]

3. Picque, N.; Haensch, T.W. Frequency comb spectroscopy. *Nat. Photonics* **2019**, *13*, 146–157. [CrossRef]
4. Coddington, I.; Swann, W.C.; Newbury, N.R. Coherent Multiheterodyne Spectroscopy Using Stabilized Optical Frequency Combs. *Phys. Rev. Lett.* **2008**, *100*, 013902. [CrossRef] [PubMed]
5. Yan, M.; Luo, P.L.; Iwakuni, K.; Millot, G.; Hänsch, T.W.; Picqué, N. Mid-infrared dual-comb spectroscopy with electro-optic modulators. *Light Sci. Appl.* **2017**, *6*, 761–768. [CrossRef]
6. Millot, G.; Pitois, S.; Yan, M.; Hovhannisyan, T.; Bendahmane, A.; Hänsch, T.W.; Picqué, N. Frequency-agile dual-comb spectroscopy. *Nat. Photonics* **2016**, *10*, 27–31. [CrossRef]
7. Yu, H.Y.; Zhou, Q.; Li, X.H.; Wang, X.H.; Kai, N. Phase-stable repetition rate multiplication of dual-comb spectroscopy based on a cascaded Mach-Zehnder interferometer. *Opt. Lett.* **2021**, *46*, 3243–3246. [CrossRef]
8. Suh, M.G.; Yang, Q.-F.; Yang, K.Y.; Yi, X.; Vahala, K.J. Microresonator soliton dual-comb spectroscopy. *Science* **2016**, *354*, 600–603. [CrossRef]
9. Luo, X.; Tuan, T.H.; Saini, T.S.; Nguyen, H.P.T.; Suzuki, T.; Ohishi, Y. Switchable dual-wavelength mode-locked fiber laser using Saganc loop mirror. *Opt. Commun.* **2020**, *463*, 125457. [CrossRef]
10. Chen, J.; Zhao, X.; Yao, Z.; Li, T.; Li, Q.; Xie, S.; Liu, J.; Zheng, Z. Dual-comb spectroscopy of methane based on a free-running Erbium-doped fiber laser. *Opt. Express* **2019**, *27*, 11406–11412. [CrossRef] [PubMed]
11. Chen, J.; Nitta, K.; Zhao, X.; Mizuno, T.; Minamikawa, T.; Hindle, F.; Zheng, Z.; Yasui, T. Adaptive-sampling near-Doppler-limited terahertz dual-comb spectroscopy with a free-running single-cavity fiber laser. *Adv. Photonics* **2020**, *2*, 036004. [CrossRef]
12. Dai, M.L.; Liu, B.W.; Shirahata, T.; Sun, X.N.; Set, S.Y.; Yamashita, S. All-polarization-maintaining, efficiently wavelength-tunable, Er-doped mode-locked fiber laser with a compact reflective Lyot filter. *Opt. Express* **2023**, *31*, 27810–27820. [CrossRef] [PubMed]
13. Dai, M.; Liu, B.; Ye, G.; Shirahata, T.; Ma, Y.; Yamaguchi, N.; Set, S.Y.; Yamashita, S. Pump-driven wavelength switching in an all-polarization-maintaining mode-locked fiber laser incorporating a CNT/PDMS saturable absorber. *Opt. Laser Technol.* **2024**, *176*, 111002. [CrossRef]
14. Zhao, X.; Li, T.; Liu, Y.; Li, Q.; Zheng, Z. Polarization-multiplexed, dual-comb all-fiber mode-locked laser. *Photon. Res.* **2018**, *6*, 853–857. [CrossRef]
15. Gu, X.; Wang, G.; Li, Y.; Gong, H.; Liang, Y.; Wu, T.; Wang, J.; Liu, Y. Polarization-multiplexed, single-cavity dual-comb fiber laser based on a birefringent crystal and a saturable absorber. *Opt. Express* **2023**, *31*, 56–64. [CrossRef]
16. Cuevas, A.R.; Kbashi, H.J.; Stoliarov, D.; Sergeyev, S. Polarization dynamics, stability and tunability of a dual-comb polarization-multiplexing ring-cavity fiber laser. *Results Phys.* **2023**, *46*, 106260. [CrossRef]
17. Zhang, C.H.; Wu, T.; He, S.; Zhang, C.Y.; Fu, B. Multiplexed dual combs in a bidirectional nanotube-mode-locked fiber laser. *Opt. Laser Technol.* **2024**, *168*, 109865. [CrossRef]
18. Zhao, X.; Hu, G.; Zhao, B.; Li, C.; Pan, Y.; Liu, Y.; Yasui, T.; Zheng, Z. Picometer-resolution dual-comb spectroscopy with a free-running fiber laser. *Opt. Express* **2016**, *24*, 21833–21845. [CrossRef]
19. Bajek, D.; Cataluna, M.A. Megahertz scan rates enabled by optical sampling by repetition-rate tuning. *Sci. Rep.* **2021**, *11*, 22995. [CrossRef]
20. Wu, Z.C.; Liu, D.M.; Fu, S.N.A.; Li, L.; Tang, M.; Zhao, L.M. Scalar-vector soliton fiber laser mode-locked by nonlinear polarization rotation. *Opt. Express* **2016**, *24*, 18764–18771. [CrossRef]
21. Wu, Z.C.; Fu, S.N.; Jiang, K.; Song, J.; Li, H.Z.; Tang, M.; Shum, P.; Liu, D.M. Switchable thulium-doped fiber laser from polarization rotation vector to scalar soliton. *Sci. Rep.* **2016**, *6*, 34844. [CrossRef]
22. Li, T.; Zhao, X.; Chen, J.; Li, Q.; Xie, S.G.; Zheng, Z. Tri-Comb and Quad-Comb Generation Based on a Multi-Dimensional Multiplexed Mode-Locked Laser. *J. Light Technol.* **2019**, *37*, 5178–5184. [CrossRef]
23. Jin, X.X.; Liu, R.Y.; Zhou, J.Y.; Li, Z.H.; Duan, Y.M.; Zhu, H.Y. Simultaneous generation of wavelength multiplexing and polarization multiplexing from a wavelength-tunable ultrafast fiber laser. *Opt. Fiber Technol.* **2024**, *84*, 103719. [CrossRef]
24. Li, S.; Shao, J.; Shen, Q.; Shen, L.; Cao, Z.; Chen, J.; Wu, Z.; Sang, M. Multiwavelength Fiber Ring Laser with Selectable Grid Spacing. *Acta Opt. Sin.* **2007**, *27*, 1802–1806.
25. Kim, R.K.; Han, Y.G. Switchable multiple lasing oscillations in an erbium-doped fiber ring laser using a single stage of a Sagnac loop mirror. *Appl. Phys. B Lasers Opt.* **2011**, *103*, 813–818. [CrossRef]
26. Pandit, N.; Noske, D.U.; Kelly, S.M.J.; Taylor, J.R. Characteristic instability of fiber loop soliton lasers. *Electron. Lett.* **1992**, *28*, 455–457. [CrossRef]
27. Man, W.S.; Tan, H.Y.; Demokan, M.S.; Wai, P.K.A.; Tang, D.Y. Mechanism of intrinsic wavelength tuning and sideband asymmetry in a passively mode-locked soliton fiber ring laser. *J. Opt. Soc. Am. B* **2000**, *17*, 28–33. [CrossRef]
28. Chen, W.C.; Chen, G.J.; Han, D.A. Spectral sideband distribution asymmetry in fiber laser. *Opt. Spectrosc.* **2012**, *113*, 222–226. [CrossRef]

29. Tang, D.Y.; Fleming, S.; Man, W.S.; Tam, H.Y.; Demokan, M.S. Subsideband generation and modulational instability lasing in a fiber soliton laser. *J. Opt. Soc. Am. B* **2001**, *18*, 1443–1450. [CrossRef]
30. Wei, W.; Qin, G.L.; Li, B.; Duan, J.F. Linear cavity erbium-doped fiber laser employing PM-PCF based Sagnac loop comb filter for wavelength tunable laser generation. *Laser Phys.* **2021**, *31*, 105104. [CrossRef]

photonics

MDPI

Article

Extending the ICESAT-2 ATLAS Lidar Capabilities to Other Planets Within Our Solar System

John J. Degnan †

Independent Research, Odenton, MD 21113, USA; jjdegnan3rd@gmail.com

† The author is a semi-retired technical consultant who previously held several research and supervisory positions at NASA Goddard Space Flight Center (1964–2003), and served as Chief Scientist at Sigma Space Corporation (2003–2019) before entering semi-retirement as a part-time consultant. He also taught a two semester graduate course in Quantum Electronics as a Distinguished Adjunct Professor of Physics at the American University in Washington DC during the period 1989 to 1993. He cofounded the International Laser Ranging Service (ILRS) in 1997 and served as its first Governing Board Chairman from 1998 to 2002. He is a Fellow of Optica (formerly the Optical Society of America) and the International Association for Geodesy, an IEEE Senior Life Member, and a member of the Sigma Pi Sigma Physics and the Drexel 100 Honor Societies.

Abstract

The ATLAS lidar on NASA's Earth-orbiting ICESat-2 satellite has operated continuously since its launch in September 2018, with no sign of degradation. Compared to previous international single-beam spaceborne lidars, which operated at a few tens of Hz, the single-photon-sensitive, six-beam ATLAS pushbroom lidar provides 60,000 surface measurements per second and has accumulated almost 3 trillion surface measurements during its six years of operation. It also features a 0.5 m^2 telescope aperture and a single, 5 Watt, frequency-doubled Nd:YAG laser generating a 10 KHz train of 1.5-nanosecond pulses at a green wavelength of 532 nm. The current paper investigates how, with minor modifications to the ATLAS lidar, this capability might be extended to other planets within our solar system. Crucial to this capability is the need to minimize the solar background seen by the lidar while simultaneously providing, for long time intervals (multiple months), an uninterrupted, modestly powered, multimegabit per second interplanetary laser communications link to a terminal in Earth orbit. The proposed solution is a pair of Earth and planetary satellites in high, parallel, quasi-synchronized orbits perpendicular to their host planet's orbital planes about the Sun. High orbits significantly reduce the time intervals over which the interplanetary communications link is blocked by their host planets. Initial establishment of the interplanetary communications link is simplified during two specific time intervals per orbit when the sunlit image of the two planets are not displaced from their actual positions ("zero point ahead angle"). In this instance, sunlit planetary images and the orbiting satellite laser beacon can be displayed on the same pixelated detector array, thereby accelerating the coalignment of the two communication terminals. Various tables in the text provide insight for each of the eight planets regarding the impact of solar distance on the worst-case Signal-to-Noise Ratio (SNR), the effect of satellite orbital height on the duration of the unblocked interplanetary communications link, and the resulting planetary surface continuity and resolution in both the along-track and cross-track directions. For planets beyond Saturn, the laser power and/or transmit/receive telescope apertures required to transmit multimegabit-per-second lidar data back to Earth are major challenges given current technology.

https://doi.org/10.3390/photonics12111048

Keywords: planetary mapping; single photon lidar; ATLAS lidar; multimegabit interplanetary laser communications; synchronized parallel Earth and planetary orbital planes; zero point-ahead terminal acquisitions; pixilated single beam lidars for dense alongtrack and crosstrack mapping

1. Introduction

Various spaceborne multiphoton lidars, typically operating at tens of pulses per second, have been successfully developed by NASA and other international space agencies and deployed for the purpose of mapping Earth, other planets and moons, and even certain asteroids. NASA's contributions to date include the Geoscience Laser Altimeter System (GLAS) on the Earth-orbiting ICEsat-1 satellite, the Mars Orbiter Laser Altimeter (MOLA), the Mercury Laser Altimeter (MLA), and the Lunar Orbiter Laser Altimeter (LOLA). Missions to map Asteroids include the Near Laser Rangefinder, Osiris Rex, and the Japanese Hayabusa 1 and 2 missions. A recent review paper provides a summary of the goals and capabilities of spaceborne lidars to date [1].

Since the single-photon-sensitive ATLAS lidar on NASA's ICESat-2 satellite was launched on 18 September 2018 into a 500 km high orbit, it has provided well over 2 trillion topographic measurements at a rate of 60,000 per second. ATLAS has achieved this result using three "strong" and three "weak" beams, all provided by a single laser producing a 5 Watt train of 1.5-nanosecond pulses at a 10 KHZ rate, with each pulse containing only 500 microjoules of energy. The pulse energies in the strong beams were four times (125 μJ) that of the "weak" beams (31 μJ). Furthermore, the passively Q-switched and frequency-doubled Nd:YAG laser, developed by NASA Goddard Space Flight Center and its contractors, operates at a green wavelength of 532 nm and, as of the time of writing, has, remarkably, shown no signs of degradation after almost 7 years in orbit.

The original single-photon lidar concept, put forward by the current author [2] for the mapping of planets and/or their moons, had suggested 16 weak beams, resulting in 160,000 measurements per second. However, certain ICESat-2 science goals (primarily related to cryosphere, biomass, and ocean studies somewhat unique to Earth) could not be met by the weak beams alone [3]. However, if one is contemplating tens (or even hundreds) of thousands of measurements per second at a distant planet, how does one collect, store, and transmit that data back to Earth?

2. Data Collection Rates

If we assume that the range data is being downloaded to Earth via On–Off Keying (OOK), i.e., via a series of ones and zeros, each measurement requires that we record and transmit the beam number (first 4 bits for 9 to 16 beamlets) followed by n bits yielding the measured range to the surface given by

$$2^n = \frac{R_{max}}{\delta} \tag{1}$$

where R_{max} is the maximum range to the surface and δ is the desired range resolution. For the ATLAS altitude of 500 km and a range resolution of 10 cm, Equation (1) yields $n = 20$ for a total of 24 bits per measurement. We now multiply by the number of beams (16) and the laser fire rate (10 kHz) to obtain a nominal data download rate of 3.84 Megabits per second or 0.24 Megabits per second per beamlet Thus, in one Earth day, one beamlet would generate 20,736 Megabits or 20.736 Gigabits of data. Thus, in order to transfer such a large amount of data from the planet under study to Earth, the interplanetary communication system would need a roughly 4 Mbit/s capability with minimal interruptions in the data flow.

3. Gaussian Laser Beam Propagation and Optical Antennas

The lowest-order (and most useful) TEM_{00} mode of a laser has a Gaussian shape. The smallest beam radius w_0 occurs at the "waist" and is typically on the order of a few millimeters and located within the laser resonator. As we move a large distance z from the waist, the beam radius grows according to the formula

$$w(z) = w_0 \sqrt{1 + \left(\frac{\lambda z}{\pi w_0^2}\right)^2} \cong \frac{\lambda z}{\pi w_0} \tag{2}$$

where λ is the laser wavelength and z is the distance from the waist with radius w_0. Thus, far from the laser, the beam area increases quadratically with distance and is given by

$$A(z) = \pi \left(\frac{\lambda z}{\pi w_0}\right)^2 = \frac{(\lambda z)^2}{A_0} \tag{3}$$

where A_0 is the Gaussian beam area at $z = 0$. We note from the latter equation that the beam area far from the waist decreases with the square of the wavelength λ and also with increasing beam waist radius w_0. This wavelength dependence is what gives laser communications an advantage over more conventional millimeter or microwave systems that require huge antennas to achieve high data rates over interplanetary distances.

One can increase the size of the laser beam waist through a modestly sized magnifying telescope, which, in turn, can reduce the divergence of the beam in the far field by several orders of magnitude [4]. For example, prior to the turn of the century, the mm accuracy satellite laser ranging systems transmitted a few mJ, 120-picosecond pulses at rates on the order of 10 Hz via a narrow tube coaligned with a meter-class receive telescope [3]. As indicated by Equation (3), this resulted in an extremely small fraction of the light intercepting the array of retroreflectors on the satellite. However, by expanding and transmitting the beam through the receive telescope, the beam intensity at the satellite was increased by over four orders of magnitude, allowing the use of significantly lower beam energies and/or smaller telescope apertures. In addition, meter-long lasers could now be replaced by microlasers a few mm long, whose short length could result in multikHz pulse rates and orders of magnitude lower pulse energies, significantly reducing potential eye hazards for ground personnel or passengers in overflying aircraft. The author and his colleagues at NASA's Goddard Space Flight Center developed the SLR2000 system, which increased the measurement rate by over 2 orders of magnitude to 2 kHz using a 30 cm diameter transmit/receive telescope [3].

The SLR2000 design depended heavily on earlier work on optical antenna (telescope) design in support of early space laser communications efforts [4,5]. As the Gaussian laser beam is magnified, there is an optimum unobscured telescope to Gaussian beam radius ($a/w = 1.12$) that maximizes the far field intensity of the beam; however, radial truncation of the beam by the telescope aperture creates a series of very-low-intensity rings around the strong central lobe in the far field, as illustrated in the top right image of Figure 1. There is also an optimum aperture radius to beam radius ratio (a/w) that maximizes the far field gain when a secondary mirror of radius b partially obscures ($\gamma = b/a$ as in left image in Figure 1), but a greater fraction of the energy is transferred to the outer rings. Furthermore, if the telescope is defocused, the central lobe and rings merge to form a broader but lower-intensity beam, as illustrated on the right side of Figure 1.

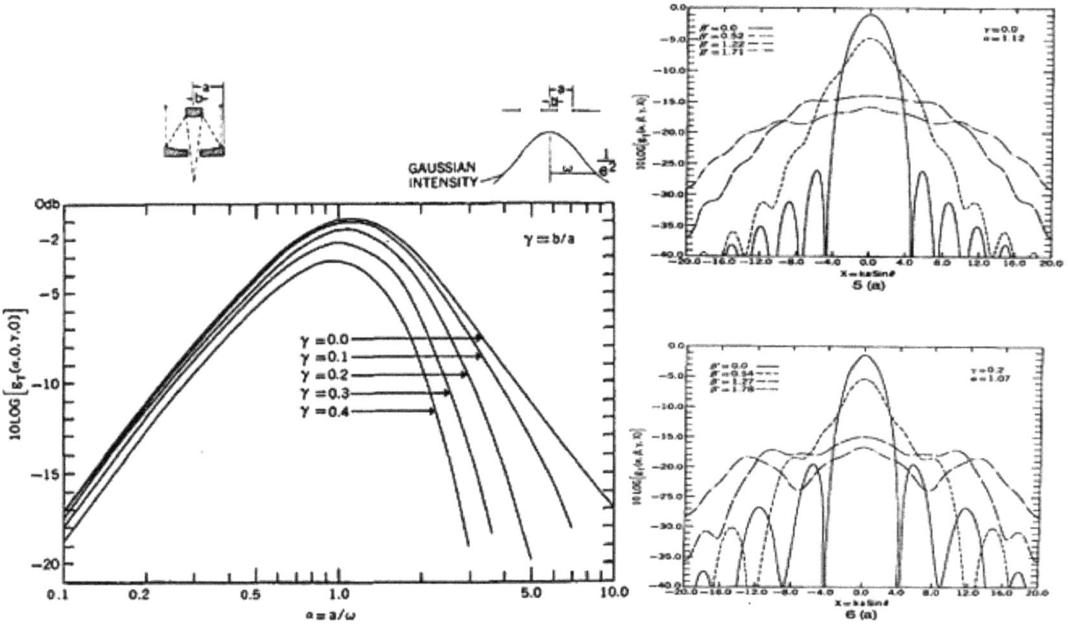

Figure 1. (**Left**) Far-field gain of a centrally obscured but focused optical telescope with a primary mirror diameter equal to 2a and a secondary mirror diameter (if any) equal to 2b as a function of the ratio $\alpha = a/w$, where w is the beam waist radius and the ratio of the secondary and primary diameters is $\gamma = b/a$. (**Upper right**) Far-field pattern of an unobscured telescope ($\gamma = 0$) without defocusing (solid line) and with two levels of defocusing (2 dashed lines). (**Lower right**) Same as upper right figure but with a secondary mirror ($\gamma = b/a = 0.2$).

4. Interplanetary Laser Communications

The present author has previously analyzed the system requirements for transmitting a given data rate via a 532 nm laser link between Earth and other planets within the solar system [6]. The results are summarized in Figure 2, which plots the laser link requirements (i.e., the transmitted laser power multiplied by the product of the transmit and receive telescope diameters squared) on the y-axis versus the distance between the Earth and the planetary terminal on the x-axis. Each of the four lines represents a different data rate, i.e.,1 Mbps, 10 Mbps, 100 Mbps, and 1 Gbps. The mean distance between Earth and the various planets within our solar system are indicated by the labelled black dots (within our solar system at the frequency-doubled Nd:YAG wavelength of 532 nm). The corresponding y values for other wavelengths λ can be obtained by simply multiplying the y-axis values in Figure 2 by $(\lambda/532\ \text{nm})^2$. Clearly, shorter wavelengths imply lower transmitted laser powers and/or smaller telescopes, but they also imply narrower beams and more stringent coalignment between collaborating telescopes.

The laser power requirements in Figure 2 assume a simple On–Off Keying (OOK) approach, i.e., a series of ones (photons detected in the bit time interval) or zeros (inadequate photons detected in the bit time interval), as described in [6]. From Figure 2, we can obtain the value of $P_T\ (D_T D_R)^2$ needed for a 16-beam, 4 Mbit/sec link at 532 nm between the various planets and Earth [2] in units of Watt-m^4, i.e., 0.04 (Mercury, Venus), 0.07 (Mars), 0.7 (Jupiter), 4.0 (Saturn), 8.0 (Uranus), 40 (Neptune), and 60 (Pluto). Of course, for the more distant planets, one could reduce the number of beams from 16 to lower the required data rates and/or place most of the link burden on the Earth receive telescope diameter D_R.

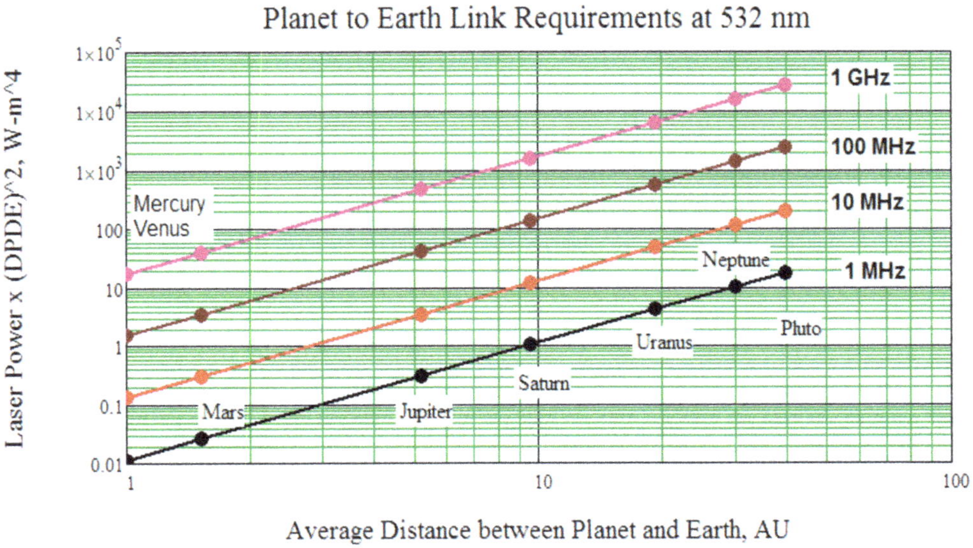

Figure 2. The product of the transmitted laser power emitted by the planetary probe and the two interacting planetary transmit (DP) and Earth receive telescope (DE) diameters squared (*y*-axis) is plotted as a function of the planetary distance (*x*-axis) for a wavelength of 532 nm and four candidate bit rates: 1 MHz (black), 10 MHz (red), 100 MHz (brown), and 1 GHz (pink). The dots along the lines indicate the average distance of the planet from Earth. The chart can be used for other wavelengths by multiplying the *y*-axis number by $\left(\lambda/532\text{ nm}\right)^2$.

5. Quasi-Continuous Communication Links via Synchronized Parallel Orbits

Quasi-continuous laser communications between Earth and the planet being mapped minimizes the buildup of stored information, the need for frequent reacquisition of interplanetary connections, and bursts of data transfers to Earth at much higher data rates. However, uninterrupted laser communications between Earth and other planets in the solar system are not possible whenever the two planets and the Sun are collinear, i.e., whenever the two planets are on opposite sides of the Sun and are therefore blocked or when they are on the same side of the Sun and the intense solar background overwhelms one of the two terminals. Nevertheless, "quasi-continuous" communications are possible if the planetary probe and Earth terminal are placed in parallel orbits perpendicular to their common orbital plane about the Sun. However, since the rotation axes of the planetary poles are not typically oriented perpendicular to the planetary orbital plane about the Sun, some areas around the poles may not be mapped without later revisions to the orbit of the mapping satellite.

In Figure 3, the maximum point-ahead angle occurs when the Earth and the planet are on opposite sides of the Sun (180°). This is due to the fact that (1) the range between them, and therefore the light travel time, is maximized, and (2) their orbital velocities are in opposite directions, and therefore, the relative velocity is additive. The point-ahead angles decrease for the outer planets because of their slower angular velocities. Smaller secondary peaks occur at 0° (or 360°) for all the planets because (1) the orbital velocities are in the same direction, and therefore, the relative velocity is reduced, and (2) the light travel time is minimized. Intermediate angles where the point-ahead angle drops to near zero would be advantageous for mutual acquisition of the two terminals and occur twice per Earth year.

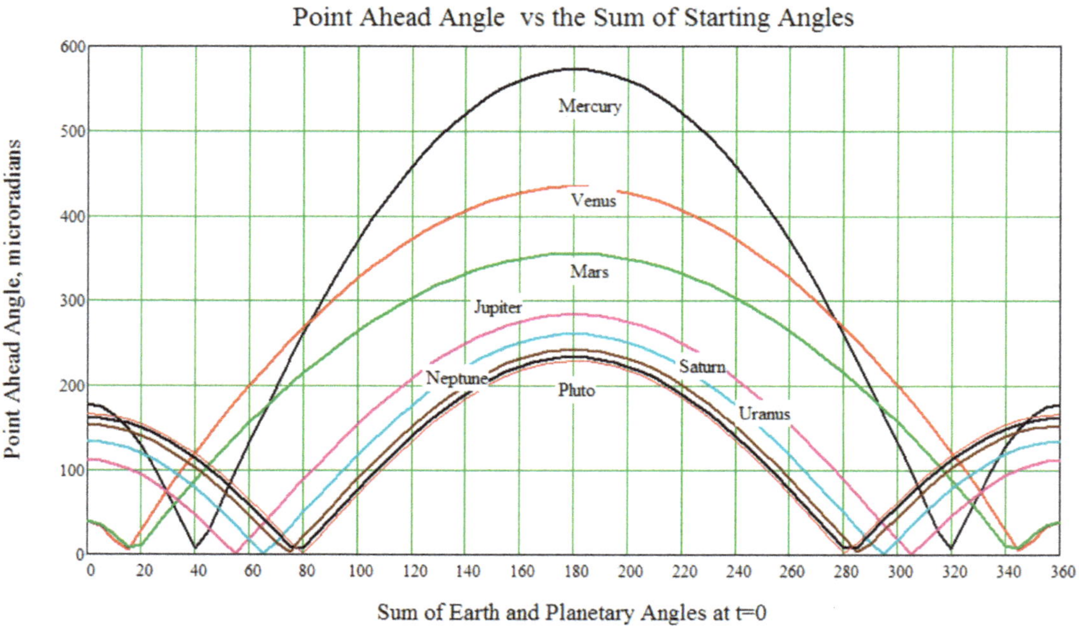

Figure 3. Terminal point-ahead angle in microradians for the various planets in the solar system when viewed from Earth versus the sum of their angular positions in their orbits about the Sun [6].

As an example, Figure 4 illustrates the manner in which an Earth–Mars link would be initially established when the angle ψ is approximately 17° and the point-ahead angle is zero, thereby enabling the Earth orbiter to use the sunlit Mars surface to initially locate the mapping satellite. The figure shows the geometry of the initial acquisition of the planetary satellite at $t = 0$ and at a later time t. The triangle in green between the Sun, Earth, and Mars has two known side dimensions corresponding to the radii of the two planetary orbits (R_E and R_M) and the opposing angle (π-ψ), thereby allowing the distance between the two satellites during acquisition to be computed and listed in Table 1 for all of the planets.

Table 1. The acquisition distances are typically not much larger than the minimum distance between Earth and the planet of interest allowing for stronger laser signals between the terminals.

Planet	Solar Distance, R_p (AU)	Acquisition Angle ψ, deg	Maximum Distance (AU)	Minimum Distance (AU)	Acquisition Distance (AU)
Mercury	0.39	40	1.39	0.61	0.66
Venus	0.72	17	1.72	0.28	0.30
Mars	1.52	17	2.52	0.52	0.54
Jupiter	5.20	55	6.20	4.20	4.56
Saturn	9.54	65	10.54	8.54	9.07
Uranus	19.20	75	20.20	18.20	19.43
Neptune	30.06	80	31.06	29.06	29.87

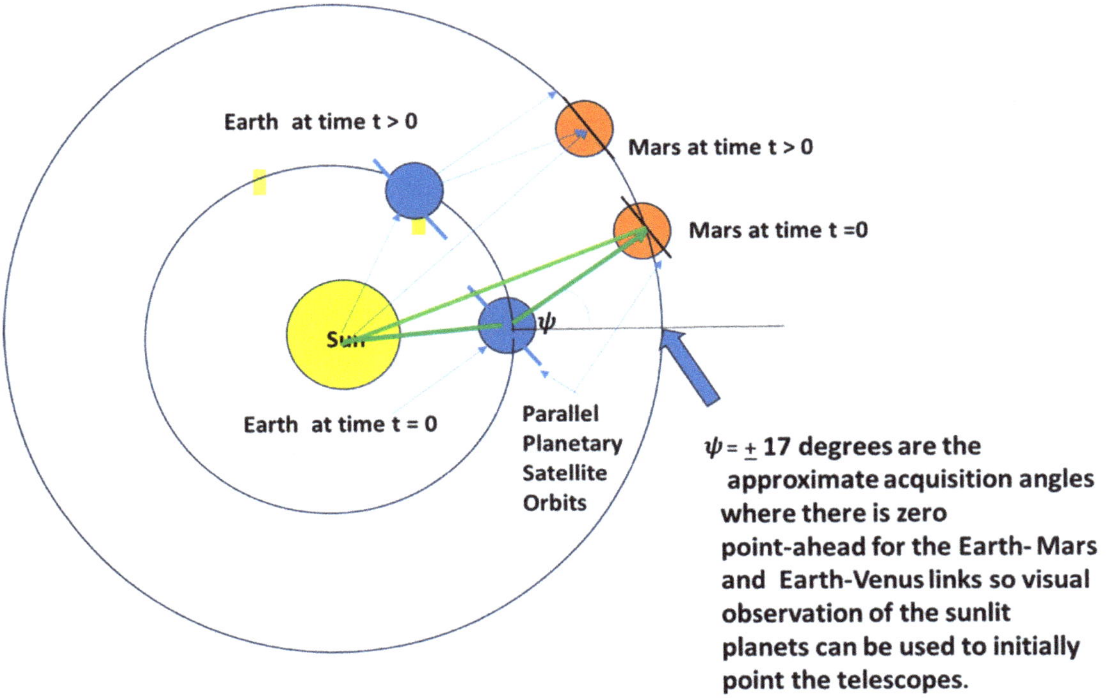

Figure 4. Earth–Mars example. Initial coalignment of the Earth terminal with the mapper at Mars takes place after the satellite pair have been placed in parallel orbits perpendicular to the common orbital plane about the Sun. When the two planets reach the geometry in the figure where the angle ψ between the Earth–Mars line and the Sun–Earth line is approximately ± 17 degrees; the point-ahead angle for both terminals is roughly zero.

At the time of acquisition, the vector between the planets is normal to the two orbits, but, as time goes on, the orbits remain parallel but the vector between them is no longer perpendicular to the orbits. Assuming circular orbits for both planets about the Sun, one can also estimate the length of time between solar interruptions of the data flow. From Figure 4, we have the following equations for the angular positions of the Earth and planetary mapping satellite

$$\theta_E(t) = \omega_E t \quad \text{(4a)}$$

$$\theta_P(t) = \omega_P t + \psi \quad \text{(4b)}$$

$$\Delta\theta_{EP} = (\omega_E - \omega_P)t - \psi \quad \text{(4c)}$$

The Sun physically blocks the beam whenever $\Delta\theta_{EP}$ is approximately equal to $n\pi$, where n is an odd integer and blinds the terminal farthest from the Sun when n is an even integer. Thus, the time interval between solar interruptions is given by

$$T = \frac{\pi + \psi}{\omega_E - \omega_P} \quad \text{(5)}$$

and the planets closest to Earth have the longest interval between solar blockages. However, the interplanetary communications link can also be blocked by the planets themselves if the satellite altitudes are too low. Furthermore, the blocking intervals are extended somewhat if the two orbits are asynchronous, as illustrated on the right side of Figure 5. Orbits are defined as synchronous if their periods and phase are equalized, as in the left side of Figure 5. Since planets have unequal masses, the satellite altitude about the planet being mapped must be chosen to match the angular speed of the Earth terminal. Thus, lidars mapping planets of lower mass than Earth will operate at a lower altitude than the Earth orbiter and vice versa.

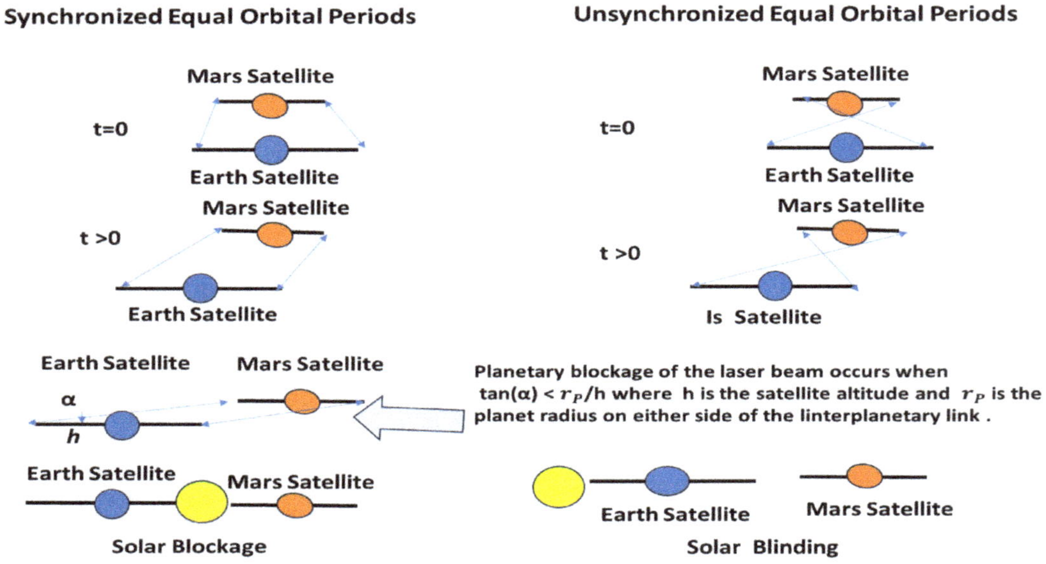

Figure 5. When both the period and phase of the two orbits are matched (synchronous orbits), the satellites always have mutual visibility, except as they (1) approach a configuration where the Sun either blocks transmission or blinds the terminal farthest from the Sun (Mars) or (2) the beams are blocked by the host planets themselves. In the latter case, blockage duration can only be reduced by increasing the Earth and planetary satellite altitudes such that they maintain the same orbital period and phase, i.e., based on planetary mass.

Figure 5 also indicates that communications between Earth and planetary satellites will not be interrupted by the host planets themselves when

$$\tan(\alpha) \geq \frac{r_P}{h} = \frac{1}{n_P} \tag{6}$$

where α is the angle that the vector makes between the two parallel orbital planes and $h = n_p r_p$ is the satellite altitude expressed as a multiple of the planetary radius r_p. If $n_p = 1$, then $\alpha = 45°$ and planetary blockage of the planet to Earth communications link occurs roughly half the time, implying the need for substantial onboard data storage and a doubling of the communications data rate when the link is unblocked. Thus, to substantially extend the measurement period without significantly increasing onboard data storage and interplanetary bit rate requirements, one must fly the Earth and planetary terminals at a much higher altitude. On the negative side, the higher orbital altitudes will (1) significantly increase the number of bits per lidar measurement to achieve the same surface height resolution, thereby resulting in greater onboard data storage and interplanetary data transmission requirements, and (2) place increased burdens on the planetary lidar link to be discussed in the next section.

6. The Lidar Link Equations for the Various Planets in Our Solar System

The ICESat-2 laser produced a 5 W beam of 1.5 nsec pulses at a rate of 10 kHz. Thus, each pulse contained 500 μJ of energy, with each photon representing 3.73×10^{-20} J at the frequency-doubled Nd:YAG laser wavelength λ = 532 nm. If we assume equal energies per beamlet, the received photons per pulse per beamlet is given by the equation

$$n_R = \frac{n_T}{n_b}(\eta_T\eta_R\eta_D)g_T g_R A_p^2 \frac{1}{(\lambda h)^2}[T_{atm}^2\rho_s(\lambda)\cos(\theta)] = \frac{3.08 \times 10^{11}}{n_b n_p^2}\frac{D^4}{(\lambda r_p)^2}[T_{atm}^2\rho_s(\lambda)\cos(\theta)] \quad (7)$$

where $n_T = 1.34 \times 10^{12}$ photons (500 μJ at 532 nm) are transmitted per pulse; n_b is the number of beamlets; the product $\eta_T\eta_R\eta_D = 0.65 \times 0.65 \times 0.7 = 0.30$ represents the typical optical efficiencies of the lidar transmit and receive optics and the single-photon-sensitive detector, respectively; g_T = 1.12 and g_R = 1 are the optimum geometric efficiencies of a common transmit/receive lidar telescope with no secondary mirror [4,5]; $A_p = \pi D^2/4$ is the area of the common transmit/receive telescope primary lens; λ = 532 nm is the ICESat-2 lidar wavelength; $h = n_p r_p$ is the lidar orbital altitude, expressed as a multiple of the planetary radius r_p; ${T_{atm}}^2$ is the two-way atmospheric transmission of the planetary atmosphere; ρ_s is the reflectance of the planetary surface illuminated by the laser; and θ is the surface slope.

The original experimental spaceborne single-photon-sensitive lidar proposed by the present author in 2002 and presented to the NASA ICESat-2 team in 2006 consisted of 16 equal beamlets generated by a single 10 kHz laser providing up to 160,000 surface measurements per second [2]. NASA originally proposed to fly the proposed system as an experiment alongside a copy of the 40 pps multiphoton, single-beam, GLAS lidar that had previously flown on ICESat-1. However, funding issues later forced the ICESAt-2 science team to choose a modified version of the 16-beam single-photon-sensitive lidar that utilized three "strong" and three "weak" beams, where the "strong" beams contained four times the energy of the "weak" beams. The need for "strong" beams was primarily driven by (1) a strong solar background when ranging over highly reflective snow and ice areas like Greenland and (2) the desire to determine tree heights within dense canopies, which provided significant attenuation of the lidar beam [3]. While the latter situations are not expected to be an issue when mapping other planets within our solar system, clouds of gases in their atmospheres and surface reflectivity will still attenuate the solar and laser radiation equally when the solar and lidar line of sight to the planetary surface coincide. However, as illustrated in Figure 5, the mapping orbits discussed previously lie in a plane normal to the planet's orbital plane about the Sun; therefore, solar/lidar path coalignment only occurs twice per planetary orbit, and only when the projection of the satellite orbital plane intersects the Sun, i.e., at or near the two solar blocking events described in Figure 5.

The solar power density entering the Earth's atmosphere at the laser wavelength of 532 nm has been shown to be about $S_E(\lambda)$ = 1.8 W/(sec-m^2-nm) = 1.8 Joule/(sec-m^2-nm) = 4.81×10^{18} photons/(sec-m^2-nm). Thus, the solar photon count at the edges of other planetary atmospheres would be given by

$$S_p(\lambda) = S_E(\lambda)\frac{R_E^2}{R_P^2} = \frac{4.81 \times 10^{18} photons}{sec - m^2 - nm}\frac{R_E^2}{R_P^2} \quad (8)$$

where R_E and R_p are the orbital radii of the Earth and planet about the Sun. Since the collinear lidar beam and solar photons see the same lidar receive telescope optics, detector, atmosphere, and surface reflectivity, the worst-case number of solar photons collected within the ATLAS laser pulsewidth (τ_p = 1.5 nsec) is given by

$$S_P(\lambda)\eta_r\eta_D g_R \tau_p [T^2_{atm}\rho_s(\lambda)\cos(\theta)] = 1.83 \times 10^6 photons \frac{R_E^2}{R_P^2} [T^2_{atm}\rho_s(\lambda)\cos(\theta)] \tag{9}$$

The worst-case solar intensity at the laser wavelength as it encounters the Earth's atmosphere is given by $S_P(\lambda)\ \Delta\lambda$, where $\Delta\lambda$ = 10 nm is the lidar filter bandwidth and R_E and R_P are the radii of the planetary orbits about the Sun. The lidar pulse is always directed normal to the nominal surface of a spherical planet but, with the satellite's orbital plane perpendicular to the planet's orbital plane about the Sun, solar rays are usually oriented at a large angle relative to the surface normal. Therefore, the vast majority of solar photons do not enter the lidar telescope until the satellite's orbital plane comes close to intercepting the Sun. Even then, only a fraction of the satellite orbit facing the Sun is subject to a strong solar background. In any event, the worst-case lidar Signal-to-Noise Ratio (SNR_{wc}) can be obtained by dividing Equation (7) by Equation (9), yielding

$$SNR \geq \frac{n_T}{n_b} \frac{(\eta_T\eta_R\eta_D) g_T g_R A_p^2 \frac{1}{(\lambda h)^2}[T^2_{atm}\rho_s(\lambda)\cos(\theta)]}{S_P(\lambda)\eta_R\eta_D g_R A_P \tau_p [T^2_{atm}\rho_s(\lambda)\cos(\theta)]} = \frac{n_T}{n_b n_P^2} \frac{(\eta_T) g_T A_p}{S_E(\lambda)\tau_p\ (\lambda r_P)^2} \frac{R_P^2}{R_E^2} = \frac{5.33 \times 10^5}{n_b n_p^2} \frac{A_P m^2}{(\lambda r_P)^2} \frac{R_P^2}{R_E^2} \tag{10}$$

High satellite altitudes ($h = n_p r_p$) positively affect the time interval over which the Earth and planetary satellite stations can maintain contact for data transfer, while negatively impacting the *SNR* and spatial resolution of the planetary lidar link due to the increased range to the surface unless one compensates with a large transmit/receive telescope. The fraction of time that interplanetary laser communications can be conducted without being interrupted by planetary blockage of the beam is given by

$$F = \frac{4\alpha}{360^\circ} \tag{11}$$

where $\tan(\alpha) = h/r_p = n_p$ and F = 0.5, 0.71, 0.80, 0.84, 0.88 for n_p = 1, 2, 3, 4, 5, respectively. Thus, for n_p = 1, the blocked and unblocked conditions have the same duration and occur alternately at 90° intervals, with the center of the first unblocked interval coinciding with the acquisition configuration illustrated in Figure 4. For n_p = 2, the angular range of unblocked intervals increases from 90° to 128° and the blocked periods are reduced from 90° to 52°. For n_p = 3, the angular range of unblocked intervals increases to 144° and the blocked periods are reduced to 36°. Clearly, however, there are diminishing returns from onboard data storage and interplanetary transfer standpoints, as n_p increases while the lidar return signal strength decreases much more rapidly (by the factor $1/n_p^2$).

Table 2 attempts to provide some insight into what a copy of the ATLAS lidar might accomplish in terms of the worst-case *SNR* for the various planets in our solar system. The relatively high worst-case *SNR* values per beamlet listed in the final two columns of the table suggest that the number of beamlets n_b can be traded against the satellite orbital height n_p (length of mapping time per orbit) or, in the case of the most distant planets (e.g., Saturn, Uranus, and Neptune), increased while still maintaining the high orbit. However, the extreme distances of the outer planets from Earth places a much larger burden on the interplanetary laser communication bandwidth, as was previously illustrated in Figure 2. On the other hand, planets farthest from the Sun also see a much lower solar background. Thus, the reduced solar noise and slower velocity of the high-altitude orbit potentially allows a greater number of beamlets and, therefore, greater resolution of the planetary topography in both the along-track and cross-track directions. The term "separate beamlets" could also refer to a simpler optical configuration in which a single ATLAS beam (1) divided into a square array of equal power beamlets generated by a diffractive device and imaged onto a pixelated, single-photon detector such as a segmented anode photomultiplier or a SPAD array, as previously utilized in high-altitude airborne lidars [7] or (2) an unmodified single-beam return imaged onto a multipixel array detector. In the latter case, the central pixel would see the maximum number of photon returns, while the surrounding pixels would see reduced but adequate photon counts because of the nominally Gaussian shape of the transmitted beam in Figure 1. As can be seen from the same figure, where the spatial intensity of the beam is plotted on a logarithmic scale in db, optimum truncation of the Gaussian beam ($\alpha = a/w = 1.12$) by the circular telescope aperture diverts a very small fraction of the laser power to the weak rings surrounding the strong central lobe. This option will be discussed further in the next section.

Table 2. The effects of planetary distance from the Sun (R_P), the radius of the planet (r_P), and the orbital height ($h = n_p r_p$), on the worst-case Signal-to-Noise Ratio (SNR_{wc}), assuming a lidar utilizing the ATLAS laser and a transmit/receive telescope with a primary area A_p = 0.5 m^2 (diameter = 0.8 m) but producing n_b beamlets and operating at much higher altitudes ($h = n_p r_p$) than the 500 km ICESat-2 orbit.

Planet	$(R_p/R_E)^2$	r_p (km)	$(\lambda r_p)^2$(m^2)	$(R_p/R_E)^2/(\lambda r_p)^2$ (m^{-4})	SNR_{wc}, A_p = 0.5 m^2 $n_b n_p{}^2 = 1$	SNR_{wc}, A_p = 0.5 m^2 $n_b n_p{}^2 = 400$
Mercury	0.152	2440	1.69	0.090	2.40×10^4	60
Venus	0.518	6052	10.37	0.050	1.33×10^4	333
Earth	1.00	6371	11.49	0.087	2.32×10^4	80
Mars	2.31	3390	3.24	0.713	1.90×10^5	475
Jupiter	27.0	66,911	1267.36	0.0213	2.87×10^4	72
Saturn	91.0	58,232	959.76	0.0948	4.31×10^5	1045
Uranus	370	25,362	182.00	0.193	3.51×10^6	880
Neptune	904	24,622	171.60	5.27	2.38×10^8	5.95×10^5

7. Planetary Surface Resolution and Continuity

The relatively high worst-case *SNR* values in the last two columns of Table 2 suggest that a large number of beamlets, or alternatively a single large area beam viewed by a large number of pixels, $n_b = n_a{}^2$ are feasible, especially at lower satellite altitudes. Low values of n_P imply (1) lower orbits and higher orbital speeds and (2) a reduction in the time available for transferring the lidar data back to Earth and a corresponding need for onboard data storage and faster interplanetary communication links. On the positive side, the low orbits permit more beamlets (or detector pixels) and therefore higher spatial resolution of the surface per pulse. Thus, one final factor that impacts our choice of the integers n_b and n_P is the spatial resolution, distribution, and continuity of the planetary terrain measurements. As the satellite altitude ($h = n_P r_p$) increases, the lidar beam area at the surface increases as n_P^2, thereby decreasing spatial resolution and lowering the *SNR* by allowing fewer lidar photons and more solar photons into the receiver. At the same time, the velocity of the satellite also decreases according to the formula

$$v_p = \sqrt{\frac{GM_P}{h}} = \sqrt{\frac{GM_P}{n_P r_P}} = 8.17 \times 10^{-6}\,\frac{m}{sec}\sqrt{\frac{M_P}{kg}\frac{km}{n_P r_P}} \tag{12}$$

where G = 6.6743 × 10^{-11} $m^3\ kg^{-1}s^{-2}$ is the Gravitational Constant, M_P is the planetary mass in kg, and $h = n_P r_P$ is the satellite altitude in km. From Equation (3), the lidar beam area on the planetary surface is approximately equal to

$$A_s \cong \frac{(n_P \lambda r_P)^2}{A_p} = \pi r_s^2 \quad (13)$$

where r_s is the radius of the circular surface area adequately illuminated by the lidar and, as before, A_p = 0.5 m^2 is the ATLAS lidar telescope primary aperture. We now assume that the lidar return is imaged onto a square detector array containing $n_b = n_a{}^2$ pixels monitoring a central square subtended by the radius r_s, as in Figure 6. This implies that

$$n_b \delta^2 = (n_a \delta)^2 = \left[2 r_s sin\left(45^\circ\right)\right]^2 = 2 r_s^2 \quad (14a)$$

or

$$n_a \delta = \sqrt{2}\, r_s = \sqrt{\frac{2}{\pi A_p}} n_P (\lambda r_P) = \frac{1.13}{m} n_P (\lambda r_P) \quad (14b)$$

where δ represents the along-track and cross-track dimensions of the square on the ground observed by a single detector pixel in the pixelated lidar detector. Table 3 displays the various values for planets in our solar system. Combined with the *SNR* results in the last column of Table 2, the results suggest that topographic resolutions in the along-track and cross-track directions at the decimeter level are quite possible, at least for the four planets closest to the Sun. For the outer planets, meter-level resolution is feasible but, as indicated by Figure 2, the interplanetary communications link becomes much more challenging due to the high laser powers and large telescope apertures needed to achieve the necessary data rates.

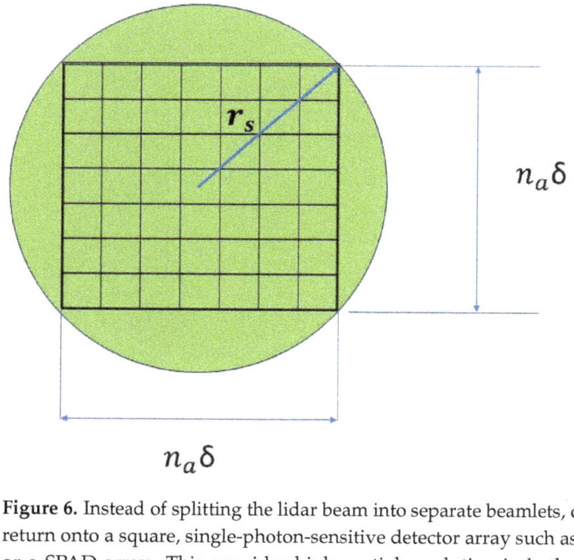

Figure 6. Instead of splitting the lidar beam into separate beamlets, one can choose to image the lidar return onto a square, single-photon-sensitive detector array such as a segmented anode MCP/PMT or a SPAD array. This provides high spatial resolution in both the along-track and cross-track dimensions. The array dimension $n_a\delta$ is chosen to fit within a radius r_s of the Gaussian beam where (1) the reduced *SNR* for the outer pixels is acceptable and (2) continuous coverage in the along-track dimension is maintained, i.e., $v_P \Delta t = n_a \delta$.

Table 3. Computation of the satellite velocities and pixelated ground dimensions for the various planets assuming the same laser power (500 mW), pulse repetition rate (10 kHz), and primary lidar telescope aperture (A_p = 0.5 m^{2r}) as the ATLAS lidar on ICESat-2. The manner in which the satellite altitude $h = n_p r_p$ affects the results is also indicated.

Planet	Planetary Mass (M_p) (kg)	Planetary Radius (r_P) (km)	$\sqrt{\frac{M_P}{kg}\frac{km}{r_P}}$	Satellite Velocity ($v_P\sqrt{n_P}$) (km/s)	(λr_p) (m^2)	Pixelated Ground-Array Dimension (m) $n_a \delta / n_p$
Mercury	3.28×10^{23}	2440	1.16×10^{10}	94.8	1.30	1.47
Venus	4.87×10^{24}	6052	2.84×10^{10}	232	3.22	3.64
Earth	5.97×10^{24}	6371	3.06×10^{10}	250	3.39	3.83
Mars	6.39×10^{23}	3390	1.37×10^{10}	112	1.80	2.03
Jupiter	1.89×10^{27}	66,911	1.68×10^{11}	1373	35.6	40.2
Saturn	5.68×10^{26}	58,232	9.87×10^{10}	806	31.0	35.0
Uranus	8.68×10^{25}	25,362	5.8×10^{10}	474	13.5	15.2
Neptune	1.02×10^{26}	24,622	6.44×10^{10}	526	13.1	14.8

8. Establishment of the Interplanetary Laser Communications Link

It is assumed that the Earth and planetary satellites are each equipped with two coaligned telescopes attached to a common pointing platform and dedicated to the interplanetary transfer of wideband planetary lidar data. One of the planetary telescopes is dedicated to transmitting 10 Megabits per second or more lidar data to the Earth terminal, while the coaligned receive telescope uses a detector array to center a beacon laser transmitted by the Earth terminal. Meanwhile, the Earth terminal receive telescope receives and stores the transmitted lidar data for later transfer to the designated Earth station(s), while the transmit telescope provides the laser beacon to optimize and maintain the optical coalignment of the two terminals. It is further assumed that the transmit telescopes can be defocused temporarily, if necessary, to allow a larger search beam during the coalignment process.

If the altitudes of the two cooperating satellites are known and initial coalignment of the two optical systems takes place when the point ahead angles between the planets are zero, as described in Figures 3 and 4, the wavelength filtered (e.g., 532 nm) sunlit image of the planet is centered on a single-photon-sensitive multipixel detector similar to that used in the pixelated lidar measurements. Laser pulses from the satellite will then be contained within a narrow ring of known radius about the full or partial planetary image within the detector array. If each terminal points its transmitter at a portion of that ring, the other terminal can fixate on the opposite terminal most quickly by scanning the ring in the opposite direction of the planetary satellite movement. However, if the two orbits are quasi-synchronized in both period and phase, as recommended previously, the signal from the opposite terminal will fall within a very small segment of the circle common to both satellites, further simplifying the coalignment. Once acquired, the beacon or lasercom beam can be centered in the array and kept there by monitoring the adjacent pixels, as is currently done in NASA's single-photon-sensitive Space Geodesy Satellite Laser (SGSLR) stations using reflections from retroreflector arrays mounted on distant Earth-orbiting satellites [8] Finally, in the event that the coalignment is temporarily lost during operations and the point ahead angle is no longer zero, defocusing the transmit telescopes enough to compensate for the known point-ahead angles in Figure 3 can aid in regaining the coalignment.

9. Concluding Remarks

While, because of cloud cover and other atmospheric conditions, conventional microwave communications systems may be best suited to collect and transmit high data rates periodically relayed by the Earth orbiter to cooperating ground stations, lasers are better suited to broadband interplanetary communications due to the λ^2 wavelength dependence mentioned in the caption to Figure 2. It is therefore assumed that, in addition to the high data rate Earth-to-ground communications terminal, the Earth-orbiting relay satellite also carries additional instrumentation in support of the interplanetary laser communications link. As illustrated in Figure 7, this would include a coaligned transmit and receive optical telescope unit and controller, a laser beacon for the planetary orbiter to lock onto, a high-data-rate interplanetary laser communications receiver, and a high-capacity data storage unit to collect and store planetary lidar data during periods when interplanetary links are blocked by the Sun or planets themselves. As discussed previously, the length of these periods depends on the choice of satellite altitudes ($h = n_P r_P$) due to planetary blocking of the signal path. For n_P =1, signal blocking occurs when the angle α in Figure 5 is less than 45°, i.e., 50% of the time, thereby doubling the required data transmission rate. While the duration of these blocking periods can be lessened somewhat by utilizing higher orbits (Equation (6)), the higher orbits have a much stronger negative impact ($1/n_P{}^2$) on both the overall lidar *SNR* (Equation (10)) and the attainable ground resolution for a given laser power and lidar telescope primary aperture A_p (Equation (14b)).

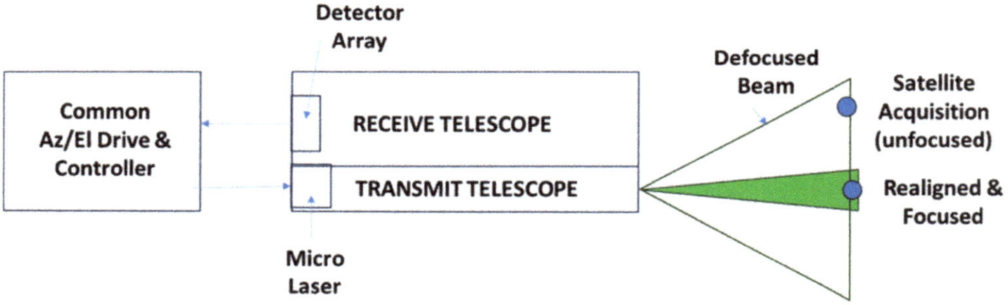

Figure 7. The Earth terminal for the wideband laser communications link consists of coaligned transmit and receive telescopes on a common gimbal mount directed by a detector array used for both initial opposite terminal acquisition and telescope reception of the wideband data. It is assumed that the transmit telescope on both terminals can be defocused to aid in the initial acquisition and coalignment of the two terminals.

It should also be noted that, according to Figure 2, the required product of laser power times the diameters of the planetary transmitting and Earth receiving telescope squared ($PD_P{}^2D_E{}^2$) increases dramatically for the planets beyond Saturn. For example, if we assume a product (DpD_E) = 1 m^2 for the Earth receive and planetary transmit telescopes and a desired data transfer rate of 10 Mbps, the laser power requirement at the 532 nm wavelength increases from 0.1 Watt for Mercury and Venus to 10 W for Saturn and 100 W for Neptune. As for the telescope diameters, most of the burden can be placed on the Earth orbiting platform. For example, NASA's 2.2 m Hubble telescope, which operated for years and was routinely serviced by astronauts in Earth orbit, would require a 45 cm diameter transmit telescope to achieve the 1 m^2 product. Increasing the planetary telescope diameter by a factor of 2 to 90 cm would reduce the laser power requirement by a factor of 4, i.e., to 2.5 W for Saturn and 25 W for Neptune.

Reducing Data Storage and Transmission Bandwidth Requirements: One could also reduce bandwidth requirements by recording and transmitting the full range recorded by the central pixel and the range difference for the remaining pixels, which would require significantly fewer bits for large arrays. For example, if we assume a 10 × 10 array (n_a = 10) containing 100 pixels, a satellite operating at an altitude four times the radius of Mars (n_P = 4 or h = 13,560 km), with a desired height resolution per pixel of 0.1 m, each pixel would require n = 24 bits ($2^n > 4h/0.1$ m = 1.3560×10^8) to record the distance to the ground, plus an additional 7 bits (2^7 = 128) to identify the pixel. Thus, 100 pixels at the ATLAS pulse rate of 10 kHz would result in a rather high data yield of 31 Mbps. However, as described in the final column of Table 3, the spacing between two adjacent pixels for our Mars example is given by

$$\delta = 2.03\ \text{m}\frac{n_p}{n_a} = 2.03\ \text{m}\frac{4}{10} = 0.812\ \text{m} = 8.12\ (10\ \text{cm}) \cong 3\ \text{bits} \tag{15}$$

where 10 cm is our desired range resolution. Thus, a vertical displacement between adjacent pixels greater than δ would imply a slope $\theta_s > 45$ degrees and an adjacent pixel return diminished by a factor of $\cos(\theta_s)$. Thus, if we retain adjacent returns (if any) differing in range by as much as 4 bits, this would imply a rather steep slope up to roughly 66 degrees between adjacent pixels. If we now only transmit the difference in range for the 99 pixels surrounding the central pixel, the bits per outer pixel would be reduced from 31 to 7 + 4 =11 bits, resulting in a 64 percent reduction in data rate, i.e., [31 + 99(11)] × 10^4 bps = 11.2 Mbps compared to 31 Mbps.

Relaying Data to an Earth Ground Station: Ultimately, the data collected by the Earth satellite via the interplanetary link must be transferred to a ground station at a high data rate. Because the Earth interplanetary terminal is orbiting in a plane orthogonal to the orbital plane of the Earth about the Sun, long-term quasi-continuous data transfer to a ground station can be accomplished via a second Earth satellite in a geosynchronous orbit, which, at an altitude of $h_g = n_g r_E$ = 35,786 km (where n_g = 5.62 Earth radii), is in constant contact with the ground station. It is interesting to note that the geosynchronous value n_g = 5.62 is comparable to the values of n_P considered in Equation (11) for the orbital synchronization of the two planetary terminals and would result in a blocking percentage between the two planetary terminals of only 10%. If also adopted for the space-to-ground link, the value n_P = 5.62 would allow the space-to-ground orbit to be synchronized with the orbits of the two planetary terminals and uninterrupted by the Earth itself; however, the desirability of this approach for a given planetary mission would require a more detailed examination.

Finally, given the orders of magnitude shorter link distances and the vulnerability of a laser link to clouds and other atmospheric effects, this space-to-ground transmission is probably best implemented using conventional microwave systems. Nevertheless, link blockage between two unsynchronized Earth-orbiting satellites will occur whenever the line between the geosynchronous satellite and the Earth terminal intersects the Earth.

Interplanetary Laser Communications Progress to Date: Almost two decades ago, using a 1.2 m telescope at Goddard Space Flight Center in Greenbelt MD, NASA demonstrated a one-way laser link to the Mars Orbiter Laser Altimeter (MOLA) over a distance of 80 million km [9] and a two-way link to the Messenger spacecraft en route to Mercury at a distance of 22 million km [10]. In 2016, a high-bandwidth, free-space laser communication link was established between Earth and the Moon by NASA's Lunar Laser Communications Development (LLCD) program. LLCD demonstrated a data downlink of 622 Mbps and an error-free downlink of 20 Mbps [11]. Even more recently, in December 2023, wideband laser communications links were successfully established between multiple ground-based Earth stations and the NASA Psyche spacecraft at Mars. Data transfer rates up to 267 Mbps over distances of 307 million miles were achieved as part of NASA's

Deep Space Optical Communications Experiment (DSOC). Since the Earth terminals were ground-based, Earth's rotation about its polar axis and/or atmospheric conditions periodically interrupted the laser link but, over time, a total pf 136 Terabits of data was successfully transferred [12]. Shortly thereafter, the European Space Agency (ESA) also successfully established a transmission–reception optical link with NASA's DSOC experiment onboard its Psyche spacecraft, located 265 million kilometers away, using two ground stations in Greece [13]. For the most recent review of deep space optical communications technology citing 247 articles, the reader is referred to a paper published in 2025 [14].

Funding: This research received no external funding.

Data Availability Statement: No inaccessible external data was used in the generation of this document.

Conflicts of Interest: Author declares no conflict of interest.

References

1. Bolcek, J.; Barakat, M.; Gibril, A.; Veverka, J.; Sloboda, J.; Marsalek, R.; Slobod, S.; Gotthans, T. Spaceborne LiDAR Systems: Evolution, Capabilities, and Challenges. *Sensors* **2025**, *25*, 3696. [CrossRef] [PubMed]
2. Degnan, J.J. Photon Counting Multkilohertz Microlaser Altimeters for Airborne and Spaceborne Topographic Measurements. *J. Geodyn.* **2002**, *34*, 503. [CrossRef]
3. Degnan, J.J. Evolution of Single Photon Lidar: From Satellite Laser Ranging to Airborne Experiments to ICESat-2. *Photonics* **2024**, *11*, 924. [CrossRef]
4. Klein, B.J.; Degnan, J.J. Optical Antenna Gain 1: Transmitting Antennas. *Appl. Opt.* **1974**, *13*, 2134–2141. [CrossRef] [PubMed]
5. Degnan, J.J.; Klein, B.J. Optical Antenna Gain 2: Receiving Antennas. *Appl. Opt.* **1974**, *13*, 2397–2401, Erratum in *Appl. Opt.* **1974**, *13*, 2762. [CrossRef] [PubMed]
6. Degnan, J.J. Multipurpose Laser Instrument for Interplanetary Ranging, Time Transfer, and Wideband Communications. *Photonics* **2023**, *10*, 98. [CrossRef]
7. Degnan, J.J. Scanning, Multibeam, Single Photon Lidars for Rapid, Large Scale, High Resolution, Topographic, and Bathymetric Mapping. *Remote Sens.* **2016**, *8*, 958. [CrossRef]
8. McGarry, J.F.; Hoffman, E.D.; Degnan, J.J.; Cheek, J.W.; Clarke, C.B.; Diegel, I.F.; Donovan, H.L.; Horvath, J.E.; Marzouk, M.; Nelson, A.R.; et al. NASA's satellite laser ranging systems for the twenty-first century. *J. Geodesy* **2018**, *93*, 2249. [CrossRef] [PubMed]
9. Abshire, J.B.; Sun, X.; Neumann, G.A.; McGarry, J.F.; Zagwodzki, T.W.; Jester, P.; Riris, H.; Zuber, M.; Smith, D. Laser Pulses from Earth Detected at Mars. In Proceedings of the 2006 Conference on Lasers and Electro-Optics and 2006 Quantum Electronics and Laser Science Conference, Long Beach, CA, USA, 21–26 May 2006; Optica Publishing Group: Washington, DC, USA, 2006; pp. 1–2.
10. Smith, D.E.; Zuber, M.T.; Sun, X.; Neumann, G.A.; Cavanaugh, J.F.; McGarry, J.F.; Zagwodzki, T.W. Two Way Laser Link over Interplanetary Distance. *Science* **2006**, *311*, 53. [CrossRef] [PubMed]
11. Cornwell, D. Space-based Laser Communications Break Threshold. *Opt. Photonics News* **2016**, *27*, 24–31. [CrossRef]
12. Available online: https://www.reddit.com/r/space/comments/1nkrt9v/nasas_deep_space_communications_demo_exceeds/ (accessed on 25 September 2025).
13. Available online: https://www.esa.int/Enabling_Support/Operations/Europe_s_first_deep-space_optical_communication_link (accessed on 25 September 2025).
14. Karmous, S.; Adem, N.; Atiquzzaman, M.; Samarakoon, S. How Can Optical Communications Shape the Future of Deep Space Communications? A Survey. *IEEE Commun. Surv. Tutor.* **2025**, *27*, 725–747. [CrossRef]

photonics

MDPI

Article

Tunable-Charge Optical Vortices Through Edge Diffraction of a High-Order Hermit-Gaussian Mode Laser

Shuaichen Li †, Yiyang Zhang †, Ying Li, Linge Mao, Pengfan Zhao and Zhen Qiao *

School of Optical-Electrical and Computer Engineering, University of Shanghai for Science and Technology, No. 516 Jun Gong Road, Shanghai 200093, China; 243350578@st.usst.edu.cn (S.L.); 2335060829@st.usst.edu.cn (Y.Z.); 242200398@st.usst.edu.cn (Y.L.); maolinge@st.usst.edu.cn (L.M.); 242200399@st.usst.edu.cn (P.Z.)

* Correspondence: qiaoz@usst.edu.cn

† These authors contribute equally to this work.

Abstract

An optical vortex is a typical structured light field characterized by a helical wavefront and a central phase singularity. With its expanding applications in modern information technology, the demand for generating vortex beams with diverse topological charges continues to grow. Existing methods for modulating the topological charges of vortex beams involve complex operations and high costs. This study proposes a novel approach to modulate the topological charges of optical vortices through edge diffraction of a high-order Hermit–Gaussian (HG) mode laser. First, a high-order HG mode laser is built using off-axis pumping configuration. By selectively obscuring specific lobes of the high-order HG beam, various optical vortices are generated using a cylindrical lens mode converter. The topological charge can be continuously tuned by controlling the number of obscured lobes. This method substantially improves the efficiency of topological charge modulation, while also enabling the generation of fractional vortex states. These advancements show potential in mode-division-multiplexed optical communications and encryption.

Keywords: optical vortex; tunable topological charge; Hermite-Gaussian mode laser; edge diffraction

1. Introduction

An optical vortex with a quantized topological charge is a typical structured light beam that carries orbital angular momentum (OAM) [1], characterized by a helical wavefront [2] and a central phase singularity. Owing to this unique property, vortex beams [3] have found expanding applications in optical communications [4,5], optical tweezers [6–9], imaging [10–12], sensing [13,14], and other fields [15], where the manipulation of topological charges [16] has played critical roles. For instance, in optical communications, employing vortex beams with various topological charges as independent channels for parallel data transmission can enhance the capacity by tens to hundreds of times [5]. In holography, vortex beams with distinct topological charges provide independent information channels, enabling multidimensional and high-security information encryption [17,18]. Therefore, the generation of diverse high-order vortex beams holds significant potential in modern information technology.

Currently, the topological charges of optical vortices are mainly manipulated [19] using extra-cavity or intra-cavity methods. Extra-cavity-based methods predominantly employ spatial phase elements, such as a spatial light modulator (SLM) [20] or a spiral phase plate (SPP) [2]. SLMs are characterized by high manufacturing costs and relatively slow response speeds due to the intrinsic response time of liquid crystal molecules. SPPs generate optical vortices by controlling the optical path difference through varying material thickness. This fundamental principle dictates that a single SPP can only generate a single vortex beam with a fixed topological charge [21,22]. Alerting the topological charge requires physical replacement of the element. By tailoring the cavity structure, laser transverse modes with controllable orders can also be directly generated within the resonator [23]. Typical methods include off-axis pumping [24] and etching patterns on cavity mirrors [25]. Off-axis pumping technology enables continuous modulation of the topological charges of vortex beams. Nevertheless, this approach requires precise control of the pump size and position, which limits its flexibility. As for the mirror surface etching technique, the topological charge is adjusted by physically moving the cavity mirror to align with a different etching spot. This operation is difficult to achieve continuous modulation of the topological charges.

In this work, we propose a simple method for tuning the topological charges of optical vortices via edge diffraction using a fixed high-order Hermit–Gaussian (HG) mode laser. Firstly, a high-order HG beam is generated directly from a laser through the off-axis pumping method. Subsequently, by selectively obscuring specific lobes of the high-order HG beam using an aperture, optical vortices with various topological charges are produced using a single-cylindrical-lens mode converter. The topological charge can be continuously tuned by adjusting the number of obscured lobes. This work employs a fixed laser output mode to generate optical vortices with tunable orders in a simple and efficient manner, significantly improving the flexibility and efficiency of topological charge modulation.

2. Materials and Methods

2.1. Experimental Setup

To generate optical vortices with continuously tunable topological charges, an experimental setup was constructed, as illustrated in Figure 1. A high-order HG mode laser was built using an off-axis pumping scheme. In detail, an 808 nm pump beam was collimated by lens L_1 and focused through lens L_2 into an Nd: YVO_4 crystal. The focal lengths of L_1 and L_2 were 100 mm and 125 mm, respectively. As shown in Figure 1a, the resonator consists of a concave mirror M_1 with a curvature radius of -100 mm, an output coupler (OC) with a transmittance of 2%, and a Nd: YVO_4 crystal with $3 \times 3 \times 5$ mm^3 and a doping concentration of 0.5%. By tilting M_1 to achieve off-axis pumping, a high-order HG beam was generated. Firstly, the HG mode was converted to an optical vortex using a single-cylindrical-lens mode converter. Thereafter, a single-side aperture was introduced to obscure a specific number of lobes of the HG beam to facilitate a lower-order transverse mode. Subsequently, a lower-charge optical vortex was obtained. In order to characterize the spatial phase of an optical vortex, a Mach–Zehnder interferometer was implemented by splitting the output laser beam into two paths [26]. In the upper path, an optical vortex was obtained using a single-cylindrical-lens mode converter. In the bottom path, one lobe of the high-order HG mode beam was used to interfere with the optical vortex. In the two paths, several lenses were used to control the beam sizes, in which the focal lengths of L_3, L_4, L_5, L_6, L_7, and L_8 were 300 mm, 500 mm, 50 mm, 300 mm, 500 mm, and 50 mm, respectively. In addition, the focal length of the cylindrical lens (CL) was 50 mm. The intensity distributions of HG modes, vortex modes, and interference patterns were recorded using a CCD camera.

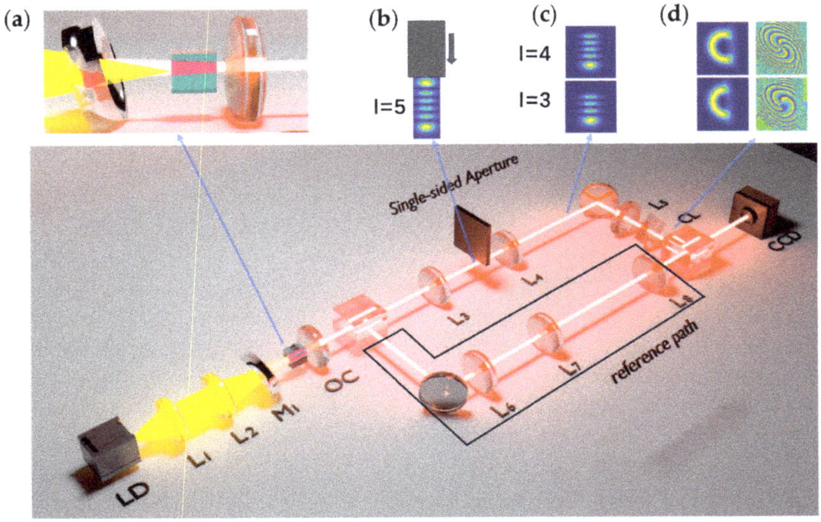

Figure 1. Schematic of the optical experimental setup. LD is an 808 nm fiber-coupled laser diode, the transmittance of OC is 2%, CL is a cylindrical lens with a focal length of 50 mm, and CCD is a camera used to record laser patterns. Blue solid frame denotes the reference path in the Mach–Zehnder interferometer. (**a**) Schematic illustration of off-axis pumping configuration. (**b**) $HG_{0,5}$ beam generated from the laser. (**c**) $TEM_{0,4}$ and $TEM_{0,3}$ beams modulated from the $HG_{0,5}$ mode. (**d**) Intensity and phase distributions of the converted optical vortices with l = 4 and l = 3 from the modulated $TEM_{0,4}$ and $TEM_{0,3}$ beams, respectively.

2.2. Simulations

In this work, optical vortices with various topological charges were obtained via edge diffraction using a fixed high-order HG mode beam, where an aperture was used to adjust the topological charges and a cylindrical lens was used to generate optical vortices. We employed the diffraction integral method to simulate the processes. In detail, the field distribution of a diffracted obscured HG mode can be expressed as follows:

$$U_{\text{in}}(x,y;z) = \frac{e^{ikz}}{i\lambda z}\iint U_0(x',y')\exp\left(\frac{ik}{2z}\left((x-x')^2+(y-y')^2\right)\right)dx'dy', \quad (1)$$

where the obscured HG mode can be described as the following formula:

$$\begin{aligned} U_0(x',y') = E_0\frac{w_0}{w(z_0)}H_m\left(\frac{\sqrt{2}x'}{w(z_0)}\right)H_n\left(\frac{\sqrt{2}y'}{w(z_0)}\right)\exp\left[-\frac{x'^2+y'^2}{w^2(z_0)} - i\frac{k(x'^2+y'^2)}{2R(z_0)}\right], \\ \times\exp\left[i\psi_{m,n}(z_0)\right]e^{-ikz_0}A(x'-x_c) \end{aligned} \quad (2)$$

In Equation (1), $U_{in}(x,y;z)$ is the complex amplitude at the front surface of the cylindrical lens; z is the propagation distance; $U_0(x',y')$ is the complex amplitude after aperture modulation; k is the wavenumber; and λ is the wavelength. In Equation (2), z_0 is the axial distance from the beam waist to the aperture; E_0 is the overall amplitude coefficient; w_0 is the waist radius; $w(z_0) = w_0\sqrt{1+(z_0/z_R)^2}$ is the beam radius at z_0; $H_m\left[\frac{\sqrt{2}x'}{w(z_0)}\right]$ and $H_n\left[\frac{\sqrt{2}y'}{w(z_0)}\right]$ are the Hermite polynomials of orders m and n, respectively; $z_R = \pi w_0^2/\lambda$ is the Rayleigh length, $R(z) = z\left[1+(z_R/z)^2\right]$ is the radius of curvature of the equal-phase surface, and the Gouy phase is $\psi_{m,n}(z) = (m+n+1)\arctan(z/z_R)$.

Specifically, $A(x' - x_c)$ is the Heaviside step function for a single-sided aperture, which equals 1 when $x' > x_c$ and equals 0 when $x' < x_c$.

A cylindrical lens was used to convert the obscured HG mode to an optical vortex, in which the power axis of the cylindrical lens is aligned at a 45° angle with respect to the x-axis. After conversion, the field distribution can be expressed as the following equation:

$$\begin{aligned} U_f(X,Y) &= \frac{e^{ikf}\cos\theta}{i\lambda f}\exp\left[\frac{ik}{2f}(X\cos\theta + Y\sin\theta)^2\right] \\ &\times \int\int_{-\infty}^{+\infty} U_{in}(x,y)\exp\left[-i\frac{2\pi}{\lambda f}(x\cos\theta + y\sin\theta)(X\cos\theta + Y\sin\theta)\right]dxdy \end{aligned} \quad (3)$$

In Equation (3), $U_f(X,Y)$ is the complex amplitude at the focal plane of the cylindrical lens, $\theta = 45°$, and f is the cylindrical lens focal length. Under the above conditions, an obscured HG mode is converted into an optical vortex through astigmatic transformation. In the process of mode transformation, the HG mode laser passes through a cylindrical lens and records in the far-field plane. The diffraction process satisfies the Fourier transform from spatial frequency domain to spatial domain. Thus, we utilized the Fourier transform method to simulate the diffraction.

We calculated the expectation value of OAM to quantify the topological charge. The expectation value of OAM can be expressed as follows:

$$\langle L_z \rangle = \frac{\iint U_f^*(\hat{L}_z U_f)dxdy}{\iint |U_f|^2 dxdy}. \quad (4)$$

In Equation (4), $\langle L_z \rangle$ is the expectation value of OAM in the z direction; $\hat{L}_z (= -i\hbar\partial/\partial\varnothing)$ is the OAM operator, where $\hbar$ is the reduced Planck constant; and $\varnothing$ is the azimuth angle. Consequently, the topological charge l can be obtained as follows:

$$l = \frac{Im\left(\iint U_f^*(\frac{\partial U_f}{\partial \phi})dxdy\right)}{\iint |U_f|^2 dxdy}, \quad (5)$$

where $Im()$ is the operation for extracting the imaginary part.

3. Results

First, we built a high-order HG mode laser and demonstrated the continuous modulation of the topological charges of optical vortices from $l = 2$ to $l = 1$. By tilting the concave mirror M_1 to a specific angle, a $HG_{0,2}$ mode beam was generated. Using a cylindrical lens, an optical vortex with the topological charge $l = 2$ was obtained. The intensity distribution of the optical vortex of $l = 2$ can be observed to exhibit a uniform ring shape. By selectively obscuring the top lobe of the $HG_{0,2}$ mode beam, the beam was modulated into the $TEM_{0,1}$ mode. With the CL removed, the intensity distribution of the $TEM_{0,1}$ was recorded after the beam was focused by L_5, as shown in Figure 2a. The corresponding converted optical vortex with the topological charge $l = 1$ is presented in Figure 2b. The intensity distribution of this optical vortex is non-uniform, with weak intensity on the right side. In contrast to the optical vortex with $l = 2$, the dark central spot of this beam is substantially smaller. To characterize the spatial phases, the built Mach–Zehnder interferometer was used to produce the interference between an optical vortex and a near-plane wave. The interference patterns are shown in Figure 2c. For the optical vortex with $l = 2$, the interference pattern presents two bifurcations in the center. For the optical vortex with $l = 1$, the interference pattern exhibits one bifurcation in the center. As such, the interference results demonstrated the tuning of the topological charge from $l = 2$ to $l = 1$. We also retrieve the corresponding phase distributions according to the interference patterns, which further clearly reveal the helical phases with $l = 2$ and $l = 1$, respectively.

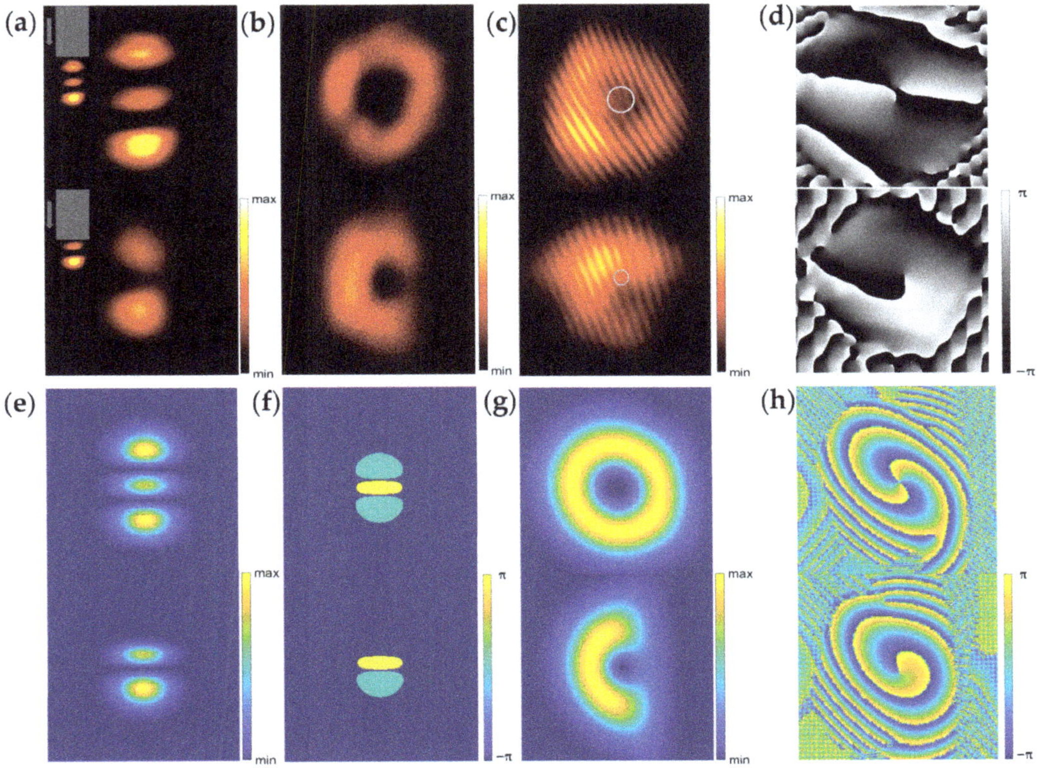

Figure 2. Topological charge modulation of optical vortices via edge diffraction of a $HG_{0,2}$ mode laser. (**a**) Laser patterns of $HG_{0,2}$ mode and the modulated $TEM_{0,1}$ mode by obscuring the top lobe of the $HG_{0,2}$ mode. Inserts show the relative positions of the aperture with respect to the $HG_{0,2}$ mode. (**b**) Laser patterns of the optical vortices converted from (**a**). (**c**) Interference patterns of the optical vortices in (**b**), with the white circles indicating the positions of fringe bifurcations. (**d**) Retrieved phase distributions of the optical vortices with $l = 2$ and $l = 1$, respectively. (**e**,**g**) Simulated intensity distributions of $HG_{0,2}$ and modulated $TEM_{0,1}$ beams, (**e**) and the corresponding converted optical vortices (**g**). (**f**,**h**) Simulated phase distributions of the $HG_{0,2}$ and $TEM_{0,1}$ beams. (**f**), and the corresponding converted optical vortices (**h**).

Numerical simulations were conducted to further reveal the mechanism for tuning the topological charges, as shown in Figure 2e–h. For the unobstructed $HG_{0,2}$ mode, the phase difference between adjacent lobes is π. This means that the phase transits twice by an amount of π across the transverse plane. As such, after conversion using a CL, the phase distribution of the converted optical vortex is characterized by two full 0 to 2π progressions per single azimuthal rotation. In contrast, when the top lobe of the $HG_{0,2}$ mode is rightly obscured, the obtained $TEM_{0,1}$ mode undergoes only single 0 to π phase shift along the transverse direction. This results in the phase distribution with only one cycle per revolution. Moreover, the simulated intensity distribution of the optical vortex with $l = 2$ exhibits a uniform ring structure, whereas that of the optical vortex with $l = 1$ is non-uniform. Both the intensity distributions and phase distributions are in good agreement with the experimental results. Therefore, these results demonstrate the feasibility of the proposed edge diffraction method for tuning topological charges.

Next, we demonstrated the modulation across an even broader range of topological charges using a higher-order HG beam laser. By tilting the concave mirror M_1 to a larger angle, we generated a $HG_{0,5}$ mode beam. Using the aforementioned edge diffraction method, a continuous transition from $HG_{0,5}$ to $TEM_{0,4}$, $TEM_{0,3}$, $TEM_{0,2}$, and $TEM_{0,1}$ was achieved and recorded by a CCD after propagating a certain distance, in which a $TEM_{0,n}$ mode has n lobes, as shown in Figure 3a. The transverse modes were then converted to optical vortices from $l = 5$, $l = 4$, $l = 3$, $l = 2$, and $l = 1$, respectively, as shown in Figure 3b. The intensity distribution of the optical vortex with $l = 5$ appears as a uniform ring, while those of the modulated optical vortices with $l = 4$, $l = 3$, $l = 2$, and $l = 1$ show non-uniform distributions. The ring diameters of optical vortices gradually decrease with the reduction in their topological charges. The observed effect is attributed to the lower orbital angular momentum (OAM) carried by a lower-order optical vortex. The diminished OAM manifests as a gentler wavefront phase variation, thereby confining the energy closer to the optical axis and resulting in a smaller ring diameter. The interference patterns of the optical vortices are shown in Figure 3c. The numbers of fringe bifurcations correspond to the topological charges of the optical vortex, which indicate that we have successfully generated optical vortices with $l = 4$, $l = 3$, $l = 2$, and $l = 1$. Thereby, the results confirmed the feasibility of the proposed method to achieve continuous modulation of topological charges over an extended range. Note that the modulated $TEM_{0,n}$ beams tend to recover to regular $HG_{0,n}$ mode with propagation, as shown in Figure 3a. This evolution can be attributed to the self-healing property of the HG mode after edge diffraction [27].

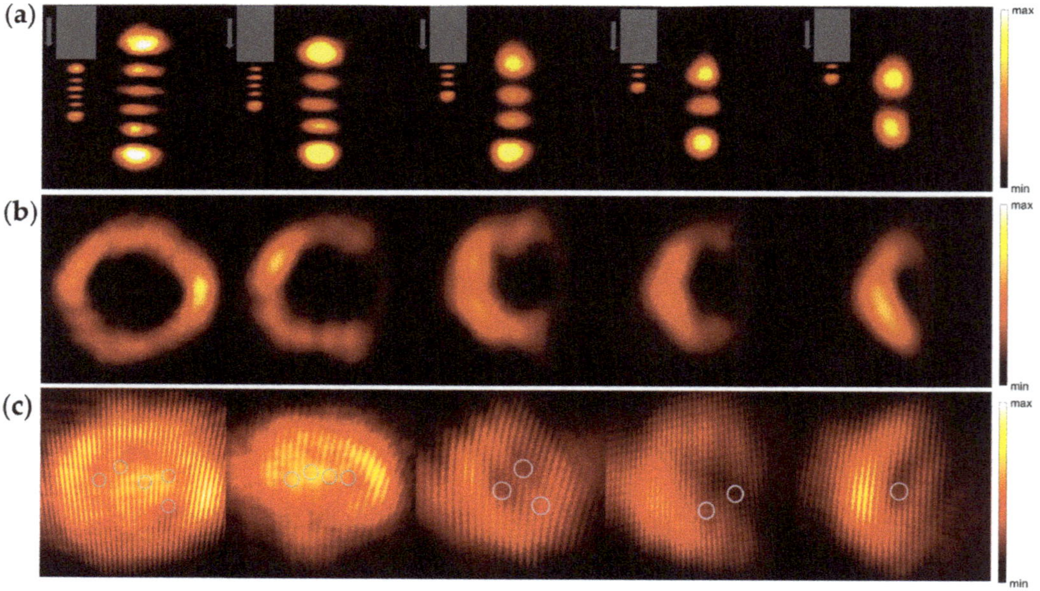

Figure 3. Topological charge modulation of optical vortices via edge diffraction of a $HG_{0,5}$ mode laser. (**a**) Laser patterns of $HG_{0,5}$ modulated $TEM_{0,4}$, $TEM_{0,3}$, $TEM_{0,2}$, and $TEM_{0,1}$ modes. Inserts show the positions of the relative aperture with respective to the $HG_{0,5}$ mode. (**b**) Laser patterns of the converted optical vortices from $l = 5$ to $l = 1$. (**c**) Interference patterns of the converted optical vortices from $l = 5$ to $l = 1$. The white circles indicate the locations of fringe bifurcations.

Furthermore, we also compared the optical vortices generated by obstructing a HG beam from different sides. A $HG_{0,5}$ mode beam was employed, and optical vortices with $l = 3$ were created by blocking two lobes from the top, bottom, and both sides simultaneously (Figure 4). The results confirm that the topological charge is determined solely by the number of obscured lobes. However, the intensity distribution of the vortex is highly sensitive to the obscuration ways: top and bottom obstructions lead to weak-intensity regions on the right and left, respectively, while dual-side obscuration reduces intensity on both sides. This demonstrates the possibility of independently controlling the topological charge and the intensity distribution of the optical vortex.

(a)
(b)
(c)

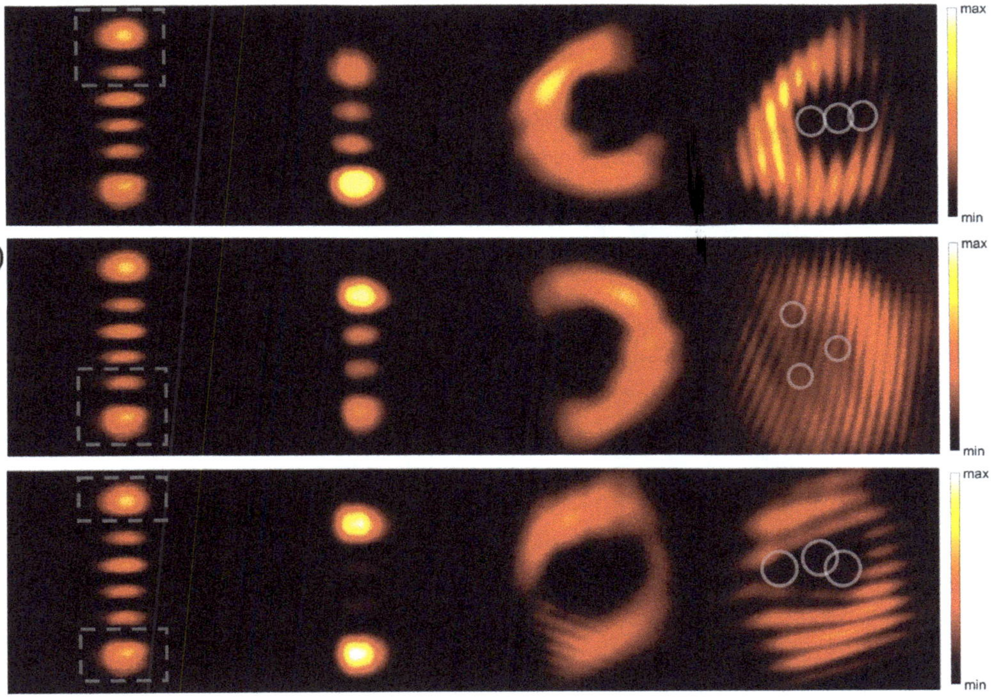

Figure 4. Experimental results of obscuring two lobes from the top, two lobes from the bottom, and one lobe from each side simultaneously in a $HG_{0,5}$ mode beam. Dotted frames on the $HG_{0,5}$ mode denote the blocking regions. The white circles indicate the locations of fringe bifurcations. (**a**) Blocking of upper two lobes: modulated $TEM_{0,3}$ mode, optical vortex, and interference patterns. (**b**) Blocking of lower two lobes: modulated $TEM_{0,3}$ mode, optical vortex, and interference patterns. (**c**) Blocking of dual-side two lobes: modulated $TEM_{0,3}$ mode, optical vortex, and interference patterns.

Finally, we recorded the optical field distributions of optical vortices under varying degrees of obscuration applied to a HG beam. By gradually obscuring the top lobe of the $HG_{0,5}$ mode, the evolution of the vortex field was recorded. We observed that when one lobe of a HG beam is partially obstructed, it generates an optical vortex with an unclosed ring, as shown in the middle column of Figure 5b. As the aperture is continuously moved until it completely blocks the entire lobe, the ring evolves from an unclosed to a closed ring profile. We hypothesize this phenomenon of an unclosed ring to be attributed to the generation of a fractional-charge vortex mode [28–30]. To verify this conjecture, we collected and analyzed its interference pattern. As shown in the central part of Figure 5c, the interference fringes exhibit distinct dislocations, which satisfy the typical features of fractional vortex beams and confirm our hypothesis. The region of fringe dislocation indicates an area of phase discontinuity, resulting in an unclosed ring. Therefore, the partially blocked state is responsible for both the generation of a fractional vortex and the non-closure of the ring. The right half of Figure 5 presents simulation results, which show good agreement with the experimental observations. The fractional vortex topological charge was calculated to be 4.23.

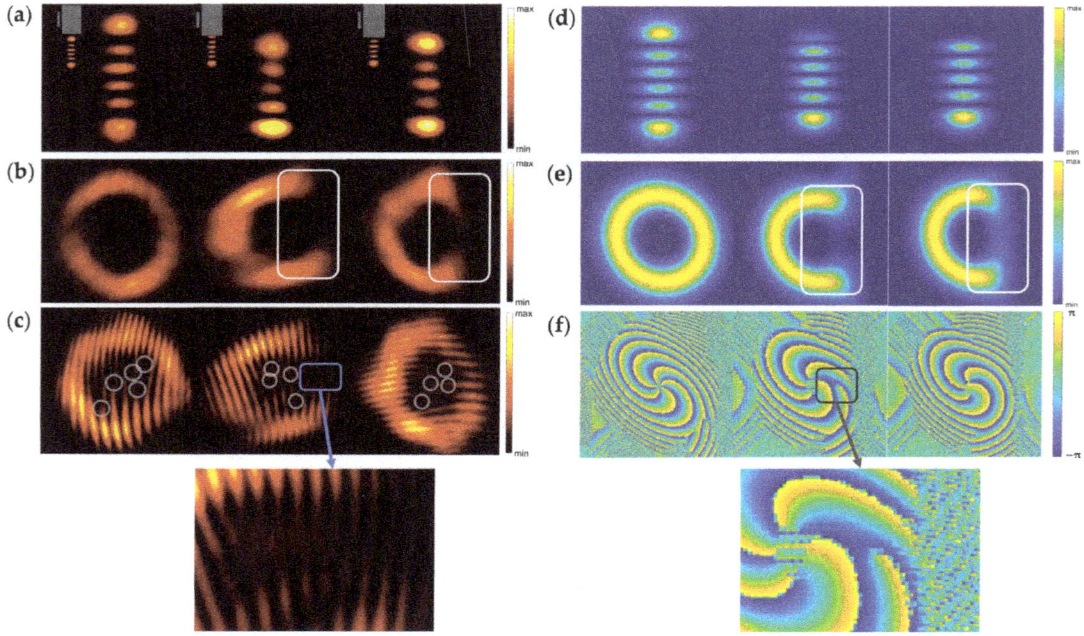

Figure 5. Experimental and simulation study of an optical vortice generation by partially obscuring a single lobe of a HG beam. (**a**) Evolution patterns from $HG_{0,5}$ to $TEM_{0,4}$. The obscuration scheme is shown in the top-left corner. (**b**) Evolution patterns of the corresponding converted optical vortices. White frames indicate the closure characteristics of the intensity distributions. (**c**) Evolution of interference patterns. Circles annotate the positions of fringe bifurcations. Inset shows dislocation details of the fringes. (**d–f**) Simulated evolution of the HG beam, corresponding optical vortices, and the phase distributions from $l = 5$ to $l = 4$. White frames highlight the ring closure of the optical vortex. Inset in (**f**) displays the enlarged phase distribution.

It is worth noting that, while the distributions of the phase singularities in both the vortices with $l = 5$ and $l = 4$ exhibit relative concentration, the phase singularities in the fractional vortex display significant dispersion (Figure 5f). As depicted in Figure 5e, the intensity distribution of the optical vortex with $l = 4$ is non-uniform yet continuous. However, the intensity distribution of the fractional optical vortex is discontinuous. We simulated the intensity at the notch position in the vortex beam as the function of obstruction position, as depicted in Figure 6. When a part of a lobe in a HG mode is blocked, the obtained optical vortex can be regarded as the coherent superposition between an integer-order vortex beam and the remaining partial lobe. As such, the destructive interference results in the intensity singularity in the ring.

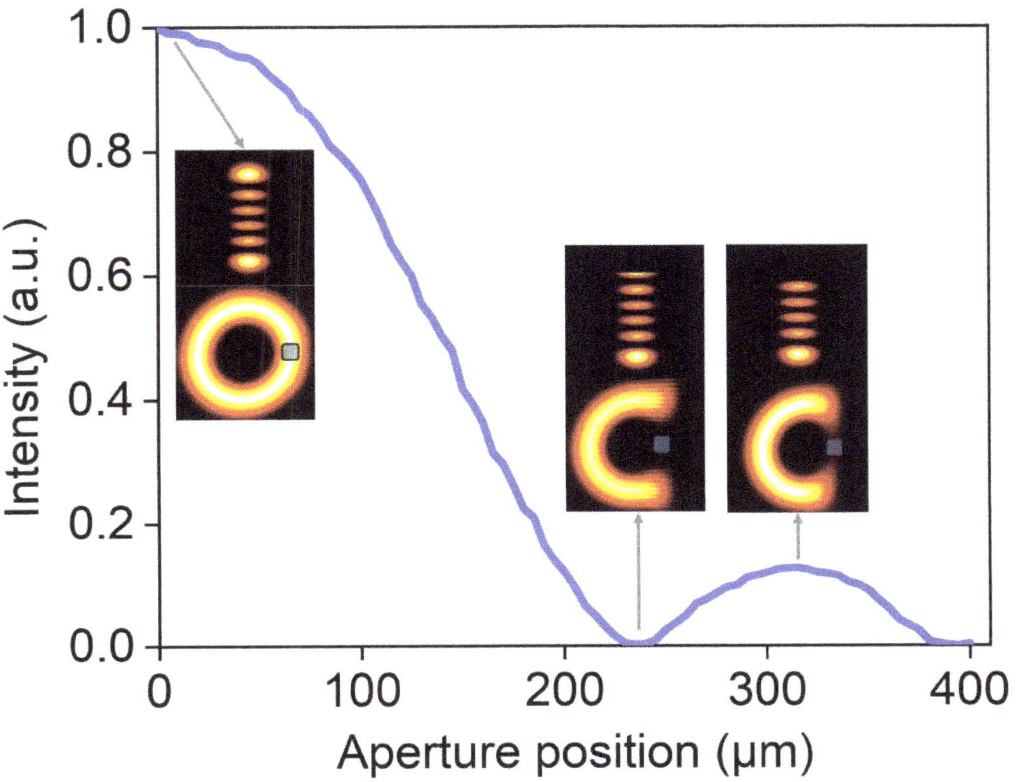

Figure 6. Intensity at the notch position in the vortex beam as the function of obstruction position. The gray squares indicate the locations of the intensity sampling points.

4. Discussion

Through the experiments, we demonstrated a dynamic method to modulate the topological charges of optical vortices by selectively blocking parts of high-order HG beams. Compared to existing methods such as SPP or cavity structure adjustment, this approach demonstrates superior flexibility and cost efficiency. Additionally, digital micromirror devices (DMDs) can operate at frequencies of several kHz when performing such tasks, while their cost is significantly lower than that of spatial light modulators (SLMs). By employing a DMD as the aperture in our system, the topological charge can be switched much faster.

In our system, the topological charge can be continuously adjusted between two adjacent integers while continuously obscuring one lobe of a HG mode. Under the current experimental conditions, the minimum step size for each adjustment is about 10 μm, and the size of one lobe is approximately 170 μm. As such, the average adjustment resolution of the topological charge is about 0.06. This capability of continuous modulation of topological charges can be applied to the coding space [31], which can significantly enhance the channel capacity of optical communication. Fractional-order vortices exhibit this missing phenomenon in their light intensity. It is equivalent to the interference between a vortex beam carrying an integer topological charge and the remaining partial lobes after obstruction. The intensity singularity is the result of destructive interference. Figure 6 shows the situation of missing intensity in fractional-order vortices. On top of that, the experiment in this study was performed based on a 1064 nm Nd: YVO_4 laser. It is worth emphasizing that the proposed method is universal and can be extended to other common laser systems, such as a Ti: Al_2O_3 laser and erbium-doped fiber laser. The mode transformation between a HG and a LG mode can also be achieved using a tilted lens [32] or an elliptically squeezed axicon [33].

The proposed method exhibits the limitation of energy loss, although the method is simple. In order to quantify the energy loss of this method, we calculated the loss rates under various blocking depths of a fixed HG_{05} mode. The energy loss is 29.44% when one lobe is blocked, 40.73% when two lobes are blocked, 50.00% when three lobes are blocked, 59.27% when four lobes are blocked, and 70.56% when five lobes are blocked. In addition, the generated vortex beams deviate from standard LG modes. This deviation arises because the obstruction of a HG mode disrupts its original spatial symmetry, which consequently breaks the spatial symmetry of the vortex beam generated through the mode conversion. Such vortex modes are difficult to apply in optical fiber communications. Nevertheless, they still hold practical potential in specific application scenarios such as optical encryption and free-space optical communications, where the encoding dimension is the topological charge itself rather than the precise intensity distribution. In addition, vortex beams with asymmetric intensity distributions also have potential applications in optical tweezers [34], where the rotating motion of a microparticle can be precisely manipulated in a local area.

5. Conclusions

This study demonstrates that introducing controllable spatial obscuration into the propagation path of a high-order HG beam enables continuous and dynamic modulation of the topological charges of optical vortices. Experimental results show that localized obscuration in the HG beam induces reconstruction of the diffraction field during propagation, thereby generating an optical vortex with the desired topological charge at the output. In addition, the intensity distribution under a fixed topological charge can also be manipulated by governing the obscuration position of the HG mode. Furthermore, by adjusting the obscuration area ratio of a single lobe, we also demonstrated the generation of a fractional optical vortex. Compared with traditional methods, the edge-diffraction-based modulation mechanism does not require complex optical alignment or phase elements, thereby reducing system complexity, enhancing modulation efficiency, and enabling dynamic control.

Author Contributions: Conceptualization, Z.Q.; methodology, Z.Q.; software, Y.Z. and S.L.; validation, S.L., Y.Z. and Y.L.; formal analysis, S.L., Y.Z., P.Z. and Z.Q.; investigation, S.L., Y.Z., Y.L. and Z.Q.; resources, Z.Q.; data curation, S.L., Y.Z. and Z.Q.; writing—original draft preparation, S.L. and Y.Z.; writing—review and editing, S.L., Y.Z., L.M. and Z.Q.; visualization, S.L., Y.Z., Y.L., L.M. and Z.Q.; supervision, Z.Q.; project administration, Z.Q. All authors have read and agreed to the published version of the manuscript.

Funding: This research was funded by the National Natural Science Foundation of China, grant number 62305216.

Institutional Review Board Statement: Not applicable.

Informed Consent Statement: Not applicable.

Data Availability Statement: The data presented in this study are available on request from the corresponding author.

Conflicts of Interest: The authors declare no conflicts of interest.

Abbreviations

The following abbreviations are used in this manuscript:

HG	Hermit-Gaussian
OAM	orbital angular momentum
OC	output coupler
CL	cylindrical lens
LD	laser diode
SLM	spatial light modulator
DMD	digital micromirror device
SPP	spiral phase plate
CCD	charge-coupled device
TEM	transverse electromagnetic mode

References

1. Wang, X.; Nie, Z.; Liang, Y.; Wang, J.; Li, T.; Jia, B. Recent advances on optical vortex generation. *Nanophotonics* **2018**, *7*, 1533–1556. [CrossRef]
2. Beijersbergen, M.W.; Coerwinkel, R.P.C.; Kristensen, M.; Woerdman, J.P. Helical-wavefront laser beams produced with a spiral phaseplate. *Opt. Commun.* **1994**, *112*, 321–327. [CrossRef]
3. Qiao, Z.; Kong, L.; Xie, G.; Qin, Z.; Yuan, P.; Qian, L.; Xu, X.; Xu, J.; Fan, D. Ultraclean femtosecond vortices from a tunable high-order transverse-mode femtosecond laser. *Opt. Lett.* **2017**, *42*, 2547–2550. [CrossRef] [PubMed]
4. Willner, A.E.; Pang, K.; Song, H.; Zou, K.; Zhou, H. Orbital angular momentum of light for communications. *Appl. Phys. Rev.* **2021**, *8*, 041312. [CrossRef]
5. Wang, J.; Yang, J.-Y.; Fazal, I.M.; Ahmed, N.; Yan, Y.; Huang, H.; Ren, Y.; Yue, Y.; Dolinar, S.; Tur, M.; et al. Terabit free-space data transmission employing orbital angular momentum multiplexing. *Nat. Photonics* **2012**, *6*, 488–496. [CrossRef]
6. Grier, D.G. A revolution in optical manipulation. *Nature* **2003**, *424*, 810–816. [CrossRef]
7. Simpson, N.B.; Dholakia, K.; Allen, L.; Padgett, M.J. Mechanical equivalence of spin and orbital angular momentum of light: An optical spanner. *Opt. Lett.* **1997**, *22*, 52–54. [CrossRef]
8. MacDonald, M.P.; Volke-Sepulveda, K.; Paterson, L.; Arlt, J.; Sibbett, W.; Dholakia, K. Revolving interference patterns for the rotation of optically trapped particles. *Opt. Commun.* **2002**, *201*, 21–28. [CrossRef]
9. Kovalev, A.A.; Kotlyar, V.V.; Porfirev, A.P. Optical trapping and moving of microparticles by using asymmetrical Laguerre–Gaussian beams. *Opt. Lett.* **2016**, *41*, 2426–2429. [CrossRef]
10. Jack, B.; Leach, J.; Romero, J.; Franke-Arnold, S.; Ritsch-Marte, M.; Barnett, S.M.; Padgett, M.J. Holographic Ghost Imaging and the Violation of a Bell Inequality. *Phys. Rev. Lett.* **2009**, *103*, 083602. [CrossRef]
11. Fürhapter, S.; Jesacher, A.; Bernet, S.; Ritsch-Marte, M. Spiral phase contrast imaging in microscopy. *Opt. Express* **2005**, *13*, 689–694. [CrossRef]
12. Li, L.; Li, F. Beating the Rayleigh limit: Orbital-angular-momentum-based super-resolution diffraction tomography. *Phys. Rev. E* **2013**, *88*, 033205. [CrossRef] [PubMed]
13. Zhu, J.; Liu, S.; Zhang, Y.-S.; Li, C.-F.; Guo, G.-C. Synthesis and observation of optical skyrmionic structure in free space. *Phys. Rev. A* **2024**, *110*, 043522. [CrossRef]
14. Emile, O.; Emile, J. Naked eye picometer resolution in a Michelson interferometer using conjugated twisted beams. *Opt. Lett.* **2017**, *42*, 354–357. [CrossRef] [PubMed]
15. Liang, H.C.; Huang, Y.J.; Lin, Y.C.; Lu, T.H.; Chen, Y.F.; Huang, K.F. Picosecond optical vortex converted from multigigahertz self-mode-locked high-order Hermite-Gaussian Nd:GdVO4 lasers. *Opt. Lett.* **2009**, *34*, 3842–3844. [CrossRef]
16. Soskin, M.S.; Gorshkov, V.N.; Vasnetsov, M.V.; Malos, J.T.; Heckenberg, N.R. Topological charge and angular momentum of light beams carrying optical vortices. *Phys. Rev. A* **1997**, *56*, 4064–4075. [CrossRef]
17. Mao, L.; Yang, Z.; Zhou, L.; Liu, J.; Su, R.; Zheng, L.; Qiao, Z.; Zhuang, S.; Zhang, D. Manipulating Elliptical Photonic Orbits in Microlasers for High-Dimensional Encryption. *Laser Photonics Rev.* **2025**, *19*, 2500238. [CrossRef]
18. Fang, X.; Ren, H.; Gu, M. Orbital angular momentum holography for high-security encryption. *Nat. Photonics* **2020**, *14*, 102–108. [CrossRef]
19. Arbabi, A.; Horie, Y.; Bagheri, M.; Faraon, A. Dielectric metasurfaces for complete control of phase and polarization with subwavelength spatial resolution and high transmission. *Nat. Nanotechnol.* **2015**, *10*, 937–943. [CrossRef]
20. Fisher, A.D. A Review of Spatial Light Modulators. In Proceedings of the Topical Meeting on Optical Computing, Incline Village, NV, USA, 18 March 1985; p. TuC1.

21. Allen, L.; Beijersbergen, M.W.; Spreeuw, R.J.C.; Woerdman, J.P. Orbital angular momentum of light and the transformation of Laguerre-Gaussian laser modes. *Phys. Rev. A* **1992**, *45*, 8185–8189. [CrossRef]
22. Coullet, P.; Gil, L.; Rocca, F. Optical vortices. *Opt. Commun.* **1989**, *73*, 403–408. [CrossRef]
23. Chen, Y.F.; Huang, T.M.; Kao, C.F.; Wang, C.L.; Wang, S.C. Generation of Hermite-Gaussian modes in fiber-coupled laser-diode end-pumped lasers. *IEEE J. Quantum Electron.* **1997**, *33*, 1025–1031. [CrossRef]
24. Lee, C.Y.; Chang, C.C.; Cho, C.Y.; Tuan, P.H.; Chen, Y.F. Generation of Higher Order Vortex Beams from a YVO4/Nd:YVO4 Self-Raman Laser via Off-Axis Pumping with Mode Converter. *IEEE J. Sel. Top. Quantum Electron.* **2015**, *21*, 318–322. [CrossRef]
25. Qiao, Z.; Xie, G.; Wu, Y.; Yuan, P.; Ma, J.; Qian, L.; Fan, D. Generating High-Charge Optical Vortices Directly from Laser Up to 288th Order. *Laser Photonics Rev.* **2018**, *12*, 1800019. [CrossRef]
26. Pezzé, L.; Smerzi, A. Phase sensitivity of a Mach-Zehnder interferometer. *Phys. Rev. A* **2006**, *73*, 011801. [CrossRef]
27. Zhao, S.; Zhang, Z.; Gao, Y.; Wang, X.; Jie, Y.; Zhao, C. Self-healing and transformation characteristics of obstructed Hermite-Gaussian modes composed structured beams. *arXiv* **2021**, arXiv:2112.03524. [CrossRef]
28. Zhang, H.; Zeng, J.; Lu, X.; Wang, Z.; Zhao, C.; Cai, Y. Review on fractional vortex beam. *Nanophotonics* **2022**, *11*, 241–273. [CrossRef]
29. Berry, M.V. Optical vortices evolving from helicoidal integer and fractional phase steps. *J. Opt. A Pure Appl. Opt.* **2004**, *6*, 259. [CrossRef]
30. Alexeyev, C.N.; Egorov, Y.A.; Volyar, A.V. Mutual transformations of fractional-order and integer-order optical vortices. *Phys. Rev. A* **2017**, *96*, 063807. [CrossRef]
31. Liu, Z.; Yan, S.; Liu, H.; Chen, X. Superhigh-Resolution Recognition of Optical Vortex Modes Assisted by a Deep-Learning Method. *Phys. Rev. Lett.* **2019**, *123*, 183902. [CrossRef]
32. Vaity, P.; Banerji, J.; Singh, R.P. Measuring the topological charge of an optical vortex by using a tilted convex lens. *Phys. Lett. A* **2013**, *377*, 1154–1156. [CrossRef]
33. Dwivedi, R.; Sharma, P.; Jaiswal, V.K.; Mehrotra, R. Elliptically squeezed axicon phase for detecting topological charge of vortex beam. *Opt. Commun.* **2021**, *485*, 126710. [CrossRef]
34. Tao, S.H.; Yuan, X.C.; Lin, J.; Peng, X.; Niu, H.B. Fractional optical vortex beam induced rotation of particles. *Opt. Express* **2005**, *13*, 7726–7731. [CrossRef]

hv photonics

MDPI

Article

154-W, Single-Frequency, Two-Stage Innoslab Amplifier at 1319 nm

Xiaochuan Zheng [1,2,3], Yanhua Lu [2,3,*], Xuguang Zhang [2,3,*], Xingwang Luo [2,3], Junzhi Ye [2,3], Peng Huang [2,3], Haoyue Shen [2,3], Tianxiang Xie [2,3], Lei Zhang [2,3], Jianli Shang [2,3], Qingsong Gao [2,3] and Weimin Wang [2,3]

1 Graduate School, China Academy of Engineering Physics, Mianyang 621900, China
2 Institute of Applied Electronics, China Academy of Engineering Physics, Mianyang 621900, China
3 National Key Laboratory on Advanced Laser and High Power Microwave, Mianyang 621900, China
* Correspondence: luyanhua@caep.cn (Y.L.); zhang_131517@163.com (X.Z.)

Abstract

A 1319 nm, single-frequency, two-stage partially end-pumped slab (Innoslab) amplifier with high output power and excellent beam quality was reported. A 3 W, quasi-continuous wave pulsed, single-frequency all-fiber seed laser was amplified to a maximum average power of 154.0 W with a magnification of ~51.3 and overall optical-to-optical efficiency up to 12.0%. The output pulse width was 132.6 μs at a repetition rate of 500 Hz. The beam quality factors of M^2 were 1.4 and 1.3 in the horizontal and vertical directions, respectively. The power stability at the maximum output power was 0.43% (RMS) in 10 min. Higher output power and optical-to-optical efficiency could be achieved through optimizing mode matching between the pump beam and the seed laser beam.

Keywords: single-frequency laser; 1319 nm; Innoslab; Nd:YAG

1. Introduction

For atmospheric adaptive optics applications in astronomy, quasi-continuous wave (QCW) pulsed sodium-guide-star lasers at 589 nm with high average power and narrow line-width have been typically employed to enhance photons returned from the sodium layer in ~90 km mesosphere while avoiding temporal overlap with backward Rayleigh scattering light in the lower atmosphere [1,2]. Up to now, the highest average output power (exceeding 100 W) of QCW pulsed sodium-guide-star lasers has been demonstrated through sum-frequency generation (SFG) of 1064 nm and 1319 nm radiation from solid-state Nd:YAG laser systems [3–5]. This configuration demands high-power outputs at both fundamental wavelengths (1064 nm and 1319 nm) with good beam quality for realizing high-power and high-efficiency SFG. However, the stimulated emission cross-section at 1319 nm is merely one-third of that at 1064 nm while the 1319 nm and 1064 nm transitions within the Nd:YAG medium share the same upper energy level [6]. This fundamental disparity leads to a lower gain and extraction efficiency while more serious thermal problems in laser resonators or amplifiers of 1319 nm. Also, fatal amplified spontaneous emission (ASE) or even parasitic oscillation of 1064 nm will cause further degradation of laser performance at 1319 nm [7–11]. These challenges make it quite difficult to achieve both high-power and high-beam-quality 1319 nm laser output of QCW pulse.

The master oscillator power amplifier (MOPA) configuration represents the standard architecture for generating high-power, narrow-linewidth laser sources of 1319 nm. Utilizing a high-quality seed source emitting at 1319 nm with high beam quality and narrow line-width, this approach enables power scaling to the hundred-watt level through multistage amplification processes [4,5]. Multiple architectures have been implemented in high-power, 1319 nm MOPA systems with QCW pulse and narrow line-width, composed of rod, zig-zag, and Innoslab [3–5,12–15]. In 2019, Yanhua Lu et al. reported a MOPA system that comprised a 0.3-GHz line-width, 1319 nm oscillator as the seed source and six Nd:YAG rod modules for laser amplification [4]. The system demonstrated an average output power of 115 W with a pulse duration of 120 μs at a repetition rate of 250 Hz. The optical-to-optical efficiency was about 9.8%, and the beam quality factors of M^2 were 3.0 in both horizontal and vertical directions. In 2017, Chuan Guo et al. implemented a QCW pulse Nd:YAG master oscillator and a multi-pass zigzag Nd:YAG slab amplifier [12]. A QCW μs-pulse, narrow line-width, 1319 nm oscillator was utilized as the seed source with an average output power of 20.5 W. An average output power of 51.5 W was achieved with a pulse width of 105 μs and a repetition rate of 500 Hz. The beam quality factors of M^2 were measured to be 1.61 and 1.81 in the orthogonal directions. Innoslab (partially end-pumped slab) laser systems employ a hybrid stable-unstable resonator configuration, effectively mitigating thermal effects to achieve both high output power and beam quality [16,17]. Because of the folded beam path, the overall gain of the amplifier and the overlap between the seed laser and pumped volume can be very high. And the good thermal management is to the benefit of the beam quality. Those innovative arrangements of the Innoslab amplifiers decrease the possibility of parasitic oscillation, guarantee the high power, high beam quality laser output with compact and reliable stability [18–23]. By employing a 10 W, narrow-linewidth, 1319 nm seed laser, a five-pass Innoslab amplifier achieved 42.3 W output with 6.5% optical conversion efficiency [13]. The beam quality factors of M^2 were 1.13 (horizontal) and 2.16 (vertical), and the pulse width was 75 μs at a repetition rate of 1 kHz. In 2023, Xuguang Zhang et al. [15] reported a seven-pass 1319 nm Nd:YAG Innoslab amplifier for a single-frequency seed laser. A maximum average output power of 80.1 W was reached first with the pulse width of 131 μs and the repetition of 500 Hz. The beam quality factors of M^2 were 1.3 in both the vertical and horizontal directions.

In this paper, a two-stage Nd:YAG Innoslab amplifier at 1319 nm with high power and high beam quality was reported for the first time. The 3 W single-frequency, QCW pulsed, all-fiber seed laser was amplified to a maximum average output power of 154.0 W, and the overall optical-to-optical efficiency was 12.0%. The average output power was nearly two times that reported before [15]. The optical-to-optical efficiency was 16.1% and 11.9% for the first- and second-stage, respectively. The output pulse width was 132.6 μs at a repetition of 500 Hz, while the beam quality factors of M^2 were 1.4 and 1.3 in the orthogonal directions, respectively.

2. Experimental Setup

The two-stage single-frequency Nd:YAG Innoslab amplifier at 1319 nm is shown schematically in Figure 1. To avoid repetitive descriptions, a clearer schematic explanation of the amplification stages and optical components can be found in reference [15]. The design parameters of the two-stage QCW pulsed amplifier are summarized in Table 1.

The first amplifier stage is composed of three parts: (1) an all-fiber single-frequency seed laser of 1319 nm, (2) two pump coupling systems of 808 nm laser-diode (LD) stack, and (3) a seven-pass hybrid cavity which comprises a Nd:YAG slab and two cylindrical mirrors. The seed laser source contained a single-frequency 15 mW CW oscillator and a QCW pulsed Raman fiber amplifier [15]. In the experiment, the central wavelength of the seed source was set at ~1319.36 nm, and the pulse width of the QCW pulsed seed was ~150 μs at a repetition of 500 Hz. The seed laser delivered 3 W maximum average power with a peak power of ~40 W. The beam quality factors of M^2 were both 1.1 in the

https://doi.org/10.3390/photonics13050449

orthogonal directions. An optical isolator was inserted between the seed laser and the Innoslab amplifier. The seed laser was shaped into an elliptical beam for better mode matching with the pump volume. The beam size after shaping was 0.84 mm and 0.48 mm along the horizontal and vertical directions, respectively, as shown in Figure 2a. The input laser peak intensity was ~1.8 times the saturation intensity of the 1319 nm laser in Nd:YAG, which pledged high extraction efficiency in the seven-pass amplification stage [15]. The seed laser was launched into the Nd:YAG slab by the 45° mirrors M1 and M2, and steered out of the cavity by the 45° mirror M3. Both M2 and M3 were high-reflective at 1319 nm and anti-reflective at 808 nm.

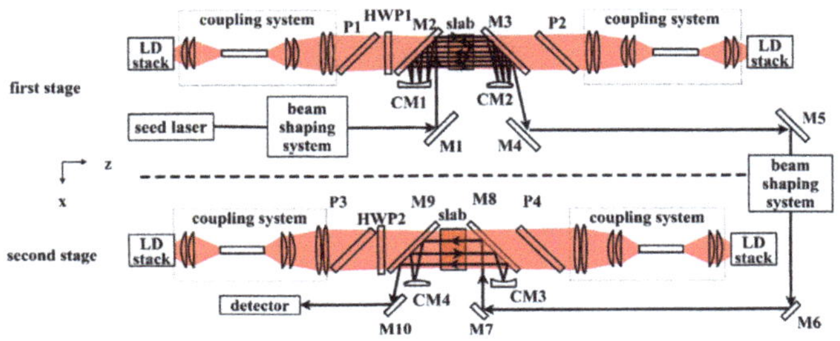

Figure 1. Experimental setup of two-stage QCW Innoslab amplifier at 1319 nm.

Table 1. Design parameters of the two-stage amplifier.

Parameters	Seed Laser	1st Stage	2nd Stage
Laser crystal	--	Nd:YAG	Nd:YAG
Nd^{3+} doping level	--	1.1 at.%	1.1 at.%
Crystal size	--	$15 \times 10 \times 1.5$ mm^3	$22 \times 10 \times 2.5$ mm^3
Pump beam size	--	14×0.65 mm^2	22×1.4 mm^2
Maximum average pump/seed power @500 Hz	3 W@150 μs (seed)	590 W@160 μs (pump)	~1050 W@160 μs (pump)
Number of passes in the crystal	--	7	3
Length of cavity	--	100 mm	150 mm
Beam quality (M^2)	1.1	1.3	1.4 (x)/1.3 (y)

The dimensions of the first-stage Nd:YAG slab were 15 mm in width, 10 mm in length, and 1.5 mm in height, which was 1.1 at.% doped and welded between two copper heat sinks (15 mm × 10 mm) for high-efficiency heat dissipation. The heat sinks were cooled by circulating water at a temperature of 25 °C. The Nd:YAG slab was mounted in the middle of the hybrid cavity, and its two optical end-facets (15 mm × 1.5 mm) were anti-reflective coated for the wavelengths of 1064 nm, 1319 nm and 808 nm. The Nd:YAG slab was dual-end-pumped by two commercial QCW pulsed LD stacks at 808 nm. Each LD stack was vertically encapsulated by ten diode bars which were collimated by a microlens in the fast-axis (i.e., y-axis) direction. By regulating the cooling water temperature, the pump center wavelength was stabilized near 808 nm. The pump radiation emitted by each stack was focused into a planar waveguide for spatial homogenization in the slow axis direction. Subsequently, it was directed out of the waveguide and imaged into the Nd:YAG slab. In the fast axis, the collimating pump beam was focused directly into the slab. The pump beam from each stack was shaped to a 14 mm×0.65 mm line-shape inside the Nd:YAG slab, as shown in Figure 2b. The intensity distribution was quasi-uniform in the horizontal direction and Gaussian in the vertical direction of the Nd:YAG slab. Two polarizers (P1 and P2) and a half-wave plate (HWP1) were inserted to protect the LDs against the resident unabsorbed pump light from the other side. The QCW LD stacks provided a maximum

https://doi.org/10.3390/photonics13050449

average pump power of 590 W (corresponding peak power of ~7.4 kW). The pulse width was ~160 μs at a repetition of 500 Hz. The overall transmittance of the pump coupling system was ~90%, and the absorption efficiency of the Nd:YAG crystal was ~90%. The maximum absorbed average pump power by the Nd:YAG crystal was approximately 478 W. The temperature of the first-stage crystal rose to 75.4 °C under the maximum pump power.

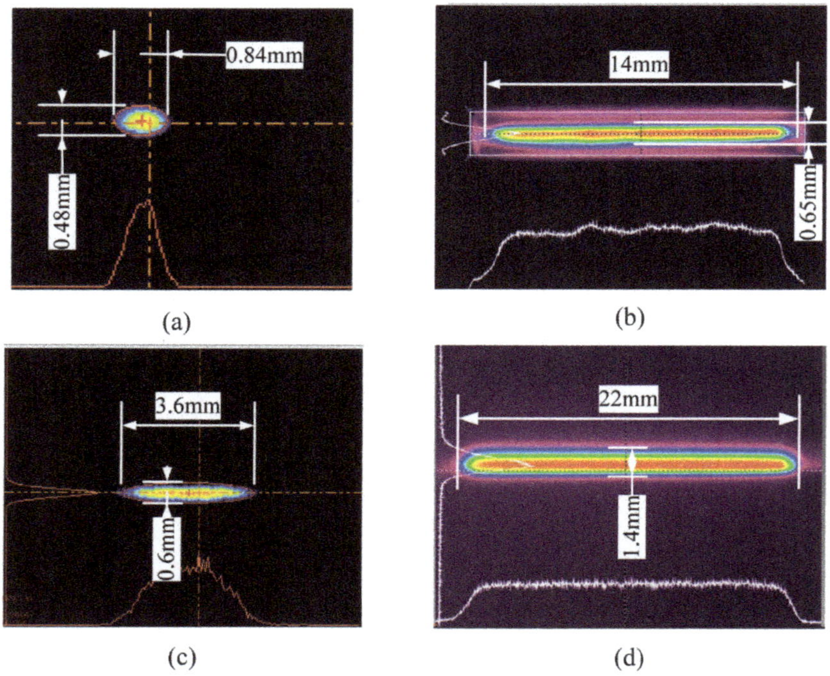

Figure 2. Beam profiles of the (**a**) first-stage input laser, (**b**) first-stage pump beam, (**c**) second-stage input laser and (**d**) second-stage pump beam.

In the first-stage amplifier, a concave cylindrical mirror CM1 (600 mm radius of curvature) and a convex cylindrical mirror CM2 (−400 mm radius of curvature) were used as the cavity mirrors. Both cavity mirrors were coated with high-reflective (HR@1319 nm) and anti-reflective (AR@1064 nm) optical coatings. The cavity mirrors were curved along the x-axis but remained planar along the y-axis. Consequently, this configuration achieved high power-extraction efficiency while maintaining the laser intensity well below the damage threshold of the optical components. Notably, the cavity exhibits stability along the y-axis direction, and the laser mode can be consistently reproduced with each passage through the gain medium, while the thermal focal length f_{th} and the cavity length L are satisfied as $L = 2f_{th}$. In the experiment, the thermal focal length of the crystal was measured as f_{th} = 60 mm at the maximum pump power. The cavity length between CM1 and CM2 was configured at ~100 mm. The mode overlapping efficiency between the pump volume and the multi-pass seed laser beam was ~74%.

The output laser was redirected from the first-stage amplifier by M4 and launched into the second-stage amplifier by three 45° mirrors, M5, M6 and M7, all of which were high-reflective coated at 1319 nm. The laser was then shaped to 3.6 mm × 0.6 mm, as shown in Figure 2c.

The second-stage amplifier adopted a three-pass configuration. The output beam from the first stage was coupled into the Nd:YAG slab via a 45° mirror (M8), then redirected

https://doi.org/10.3390/photonics13050449

outward through mirrors M9 and M10. M8 and M9 were high-reflective coated at 1319 nm and anti-reflective coated at 808 nm. A concave cylindrical mirror, CM3 (600 mm radius of curvature), and a convex cylindrical mirror, CM4 (−300 mm radius of curvature), were used as the cavity mirrors. Both of the mirrors were coated with high-reflective films at 1319 nm and anti-reflective films at 1064 nm. The size of the second-stage Nd:YAG slab was 22 mm × 10 mm × 2.5 mm. The Nd:YAG slab was 1.1 at.% doped. Similarly, the pump beam from each LD stack was shaped into a line shape of 22 mm × 1.4 mm within the Nd:YAG slab, as shown in Figure 2d. The two LD stacks provided a maximum average pump power of 1050 W (corresponding peak power of 13 kW). The pulse width was ~160 μs at a repetition of 500 Hz. The absorption efficiency of the Nd:YAG crystal was ~85%, and the maximum absorbed average pump power by the Nd:YAG crystal was approximately 803 W. At the maximum pump power, the thermal focal length of the crystal was 75 mm, and the cavity length between CM3 and CM4 was configured at ~150 mm. The temperature of the second-stage crystal increased to 84.4 °C under the maximum pump power.

3. Results and Discussion

The QCW LD pump pulse widths of the first- and second-stage amplifiers were the same, 160 μs, and the repetitions were both 500 Hz. After temporally synchronizing the seed laser and the QCW pump sources, two-stage amplification was carried out. The average output power and optical-to-optical efficiency of the first- and second-stage versus absorbed average pump power are shown in Figure 3. In the experiments, the maximum average output power of the first-stage amplifier was 80.1 W with an optical-to-optical efficiency of 16.1%, while the second-stage amplifier achieved 154.0 W with an optical-to-optical efficiency of 11.9%. The overall optical-to-optical efficiency of the two-stage amplifier was 12.0%. The high peak pump intensity generated significant gain for the 1319 nm laser, directly enhancing the amplification efficiency. Additionally, the input peak intensity (~1.8 times the saturation intensity) further guaranteed high-efficiency amplification. The maximum optical-to-optical efficiency of the first- and second-stage was 16.4% and 12.3% with the average output power of 73.9 W and 136.5 W, respectively. To the best of our knowledge, this is the first report of such high average output power and optical-to-optical efficiency for 1319 nm high-power amplifier systems [12–15]. As pump power increased, the thermal focal length along the height dimension decreased below half the cavity length, which thereby brought about decreased overlap efficiency between the seed laser beam and the pump volume. Concurrently, ASE at the main emission line of 1064 nm intensified with higher pumping. These combined effects degraded the optical-to-optical conversion efficiency at 1319 nm [14,15].

The input laser beam exhibited vertical dimensions of 0.48 mm (first stage) and 0.6 mm (second stage), while the pump beams measured 0.65 mm and 1.4 mm vertically in their respective stages. Such a mismatch in beam sizes between seed and pump radiation was more significant in the second stage [14,15]. Therefore, the overlapping efficiency of the second stage was only 51% which was lower than that of the first stage (approximately 74%). In the simulation, the average output power and optical-to-optical efficiency can be improved to over 210 W and 20.9% with an overlapping efficiency of 75%.

The elliptical output beam from the two-stage amplifier was transformed into a nearly circular beam by a cylindrical telescope. At the average output power of 154.0 W, the beam quality factors of M^2 in the horizontal and vertical directions were 1.4 and 1.3, respectively (see Figure 4). As depicted in Figure 5, the output pulse width was 132.6 μs at a repetition of 500 Hz. The peak power was above 2.3 kW. No self-excitation was observed under these high peak pump power conditions.

https://doi.org/10.3390/photonics13050449

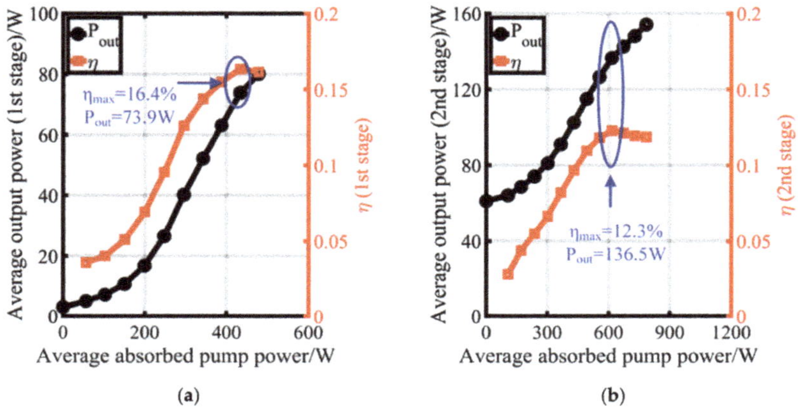

Figure 3. Average output power (Pout) and optical-optical efficiency (η) of the (**a**) first- and (**b**) second-stage versus absorbed average pump power.

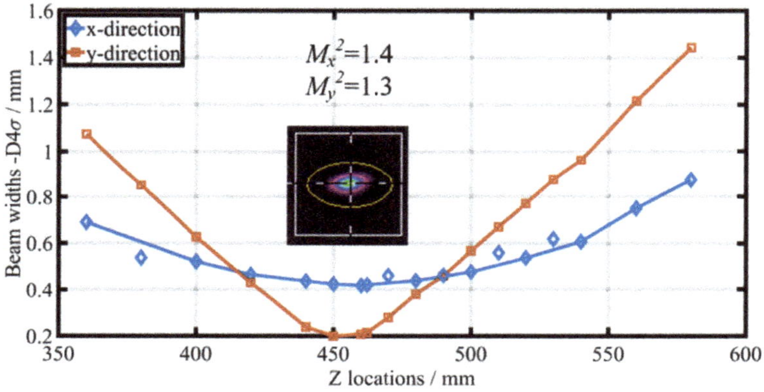

Figure 4. Beam quality measurement at the output power of 154.0 W.

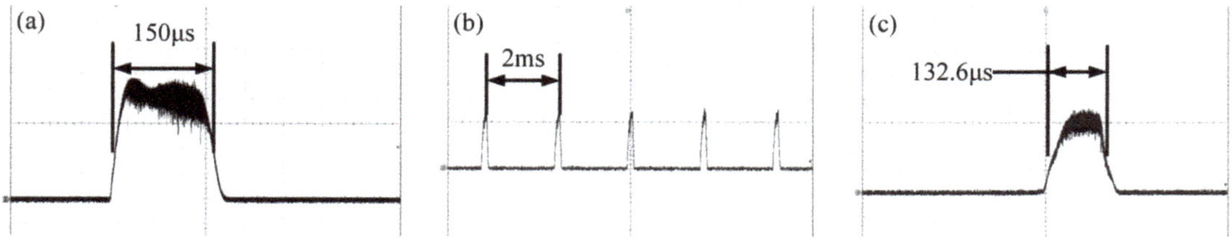

Figure 5. Pulse waveform of the seed laser (**a**), pulse train (**b**) and pulse waveform (**c**) of the 154.0 W output laser.

Furthermore, we found that the central wavelength and the line-width of the output laser were nearly the same as those of the seed laser. The central wavelength of the output laser was measured to be 1319.367 nm with a line-width of ~0.6 MHz. The power stability in 10 min was measured to be 0.43% (RMS), as shown in Figure 6. The degree of linear polarization was measured to be 15 dB after amplification while the linear polarization degree of the seed laser was 20 dB.

https://doi.org/10.3390/photonics13050449

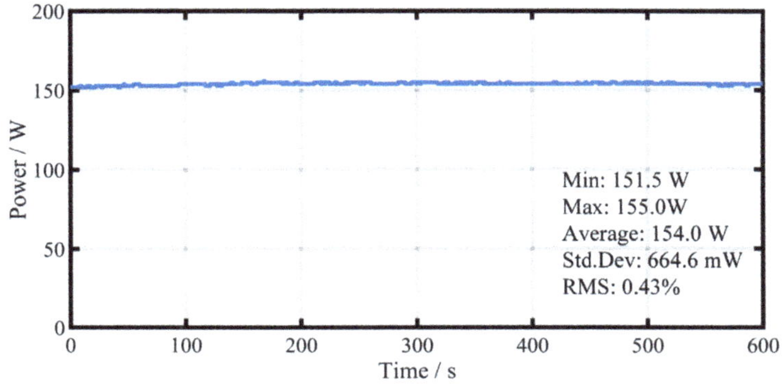

Figure 6. Power stability in 10 min at the maximum output power.

4. Conclusions

In summary, a high-power, single-frequency, QCW pulsed laser at 1319 nm was realized by a two-stage Nd:YAG Innoslab amplifier. The maximum average output power was 154.0 W with a magnification of ~51.3, and the output pulse width was 132.6 μs at a repetition of 500 Hz. At the maximum output power, the beam quality factors of M^2 were 1.4 and 1.3 in the horizontal and vertical directions, respectively. Thanks to the high peak pump intensity provided by the QCW pump source, the overall optical-to-optical efficiency of the two-stage amplifier was 12.0%. To the best of our knowledge, this is the first report of such high average output power and optical-to-optical efficiency for 1319 nm high-power amplifier systems [12–15]. With higher overlapping efficiency between the seed laser and the pump volume, the output power and optical-to-optical efficiency can be further improved. The output characteristics of the 1319 nm are conformable to achieve high power 589 nm sodium-guide-star laser by sum-frequency with the QCW single-frequency 1064 nm laser reported in related literature [22,23]. Nearly an average output power of higher than 200 W at 589 nm has been achieved for the QCW pulsed sodium-guide-star applications.

Author Contributions: Conceptualization, X.Z. (Xiaochuan Zheng), Y.L. and X.Z. (Xuguang Zhang); Methodology, X.Z. (Xiaochuan Zheng), Y.L., X.Z. (Xuguang Zhang) and X.L.; Software, X.Z. (Xiaochuan Zheng); Validation, X.Z. (Xiaochuan Zheng), X.L., J.Y., P.H., H.S., T.X. and L.Z.; Formal analysis, X.Z. (Xiaochuan Zheng); Investigation, X.Z. (Xiaochuan Zheng) and X.L.; Resources, X.Z. (Xuguang Zhang); Data curation, X.Z. (Xiaochuan Zheng); Writing—original draft, X.Z. (Xiaochuan Zheng); Writing—review & editing, Y.L. and X.Z. (Xuguang Zhang); Visualization, X.Z. (Xiaochuan Zheng); Supervision, Y.L., X.Z. (Xuguang Zhang), J.S., Q.G. and W.W.; Project administration, Y.L., X.Z. (Xuguang Zhang), J.S., Q.G. and W.W.; Funding acquisition, Y.L. and X.Z. (Xuguang Zhang). All authors have read and agreed to the published version of the manuscript.

Funding: This research was funded by the Presidential Foundation of China Academy of Engineering Physics (Grant No. YZJJLX2019015 and 2022-ZLJJ-03).

Data Availability Statement: Data underlying the results presented in this paper are not publicly available at this time but may be obtained from the authors upon reasonable request.

Conflicts of Interest: The authors declare no conflicts of interest.

References

1. Bradley, L.C. Pulse-train excitation of sodium for use as a synthetic beacon. *J. Opt. Soc. Am. B-Opt. Phys.* **1992**, *9*, 1931–1944. [CrossRef]

https://doi.org/10.3390/photonics13050449

2. Telle, J.M.; Milonni, P.W.; Hillman, P.D. Comparison of pump-laser characteristics for producing a mesospheric sodium guidestar for adaptive optical systems on large-aperture telescopes. *Proc. SPIE Int. Soc. Opt. Eng.* **1998**, *3264*, 37–42. [CrossRef]
3. Lu, Y.H.; Xie, G.; Zhang, L.; Fan, G.B.; Pang, Y.; Li, N.; Wei, B.; Gao, S.X.; Zhang, W.; Tang, C. High-energy all-solid-state sodium beacon laser with line width of 0.6 GHz. *Appl. Phys. B.* **2015**, *118*, 253–259. [CrossRef]
4. Lu, Y.; Zhang, L.; Xu, X.; Ren, H.; Chen, X.; Wei, X.; Wei, B.; Liao, Y.; Gu, J.; Liu, F.; et al. 208 W all-solid-state sodium guide star laser operated at modulated-longitudinal mode. *Opt. Express* **2019**, *27*, 20282–20289. [CrossRef]
5. Bian, Q.; Bo, Y.; Zuo, J.W.; Feng, L.; Gao, H.W.; Yuan, L.; Cui, D.F.; Peng, Q.J.; Chen, H.B.; Xu, Z.Y. High-repetition-rate 100 W level sodium beacon laser for a multi-conjugate adaptive optics system. *Opt. Lett.* **2020**, *45*, 1818–1821. [CrossRef] [PubMed]
6. Koechner, W. *Solid-State Laser Engineering, Sixth Revised and Updated Edition*; Springer: Berlin/Heidelberg, Germany, 2006.
7. Li, M.L.; Zhao, W.F.; Zhang, S.B.; Guo, L.; Hou, W.; Li, J.M.; Lin, X.C. 1.86 W cw single-frequency 1319 nm ring laser pumped at 885 nm. *Appl. Opt.* **2013**, *51*, 1241–1244. [CrossRef]
8. Lin, W.; Zhang, L.; Wang, J.; Chen, Y.; Xu, L.; Wang, D.; Zhou, T.; Tong, L.; Lu, Y.; Gao, Q. Amplification Properties of 1319 nm Nd:YAG Planar Waveguide. *Laser Optoelectron. Prog.* **2019**, *56*, 011404. [CrossRef]
9. Li, L.P.; Li, Y.J.; Song, Y.J.; Tu, W.; Zong, N.; Zhang, S.J.; Wang, Z.M.; Bo, Y.; Peng, Q.J.; Xu, Z.Y. 10.3 W diode-pumped passively mode-locked Nd:YAG laser at 1319 nm with a semiconductor saturable absorber mirror. *Laser Phys.* **2019**, *29*, 095001. [CrossRef]
10. Li, N.; Yu, P.; Lu, Y.; Zhang, L.; Xie, G.; Wang, W.; Xu, X. 800 mJ High Beam Quality All-Solid-State 1319 nm Pulsed Laser. *Chin. J. Lasers* **2013**, *40*, 0802007. [CrossRef]
11. Xu, X.; Lu, Y.; Zhang, L.; Ren, H.; Chen, X. Relaxation oscillation suppressed, narrow linewidth, high beam quality, 1319 nm long-pulsed duration laser. *Appl. Opt.* **2018**, *57*, 4692–4695. [CrossRef]
12. Guo, C.; Zuo, J.; Bian, Q.; Xu, C.; Zong, Q.; Bo, Y.; Shen, Y.; Zong, N.; Gao, H.; Lin, Y.; et al. Compact, high-power, high-beam-quality quasi-CW microsecond five-pass zigzag slab 1319 nm amplifier. *Appl. Opt.* **2017**, *56*, 3445. [CrossRef] [PubMed]
13. Zheng, J.K.; Bo, Y.; Xie, S.Y.; Zuo, J.W.; Wang, P.Y.; Guo, Y.D.; Liu, B.L.; Peng, Q.J.; Cui, D.F.; Lei, W.Q.; et al. High power quasi-continuous-wave diode-end-pumped Nd:YAG slab amplifier at 1319 nm. *Chin. Phys. Lett.* **2013**, *30*, 74202. [CrossRef]
14. Zhang, X.; He, T.; Luo, X.; Chen, X.; Zhang, L.; Xu, X.; Ren, H.; Xu, L.; Lu, Y.; Sun, J.; et al. Study of long-pulse quasicontinuous wave INNOSLAB amplifier at 1319 nm. *Opt. Eng.* **2020**, *59*, 056112. [CrossRef]
15. Zhang, X.; Ye, J.; Chen, X.; Luo, X.; Zhang, L.; Lu, Y.; Shang, J.; Gao, Q.; Wang, W. 80-W single-frequency Innoslab μs-amplifier at 1319 nm with high efficiency. *Opt. Express* **2023**, *31*, 26757–26763. [CrossRef]
16. Du, K.; Wu, N.; Xu, J.; Giesekus, J.; Loosen, P.; Poprawe, R. Partially end-pumped Nd:YAG slab laser with a hybrid resonator. *Opt. Lett.* **1998**, *23*, 370–372. [CrossRef]
17. Du, K. Unique performances and favourable applications of INNOSLAB lasers. *Proc. SPIE Int. Soc. Opt. Eng.* **2009**, *7193*, 71932B–71932B–71910. [CrossRef]
18. Russbueldt, P.; Hoffmann, D.; Höfer, M.; Löhring, J.; Luttmann, J.; Meissner, A.; Weitenberg, J.; Traub, M.; Sartorius, T.; Esser, D.; et al. Innoslab amplifiers. *IEEE J. Sel. Top. Quantum Electron.* **2014**, *21*, 447–463. [CrossRef]
19. Russbueldt, P.; Mans, T.; Rotarius, G.; Weitenberg, J.; Hoffmann, H.D.; Poprawe, A.R. 400W Yb:YAG Innoslab fs-Amplifier. *Opt. Express* **2009**, *17*, 12230–12245. [CrossRef]
20. Strotkamp, M.; Elsen, F.; Löhring, J.; Traub, M.; Hoffmann, D. Two stage Innoslab amplifier for energy scaling from 100 to >500mJ for future lidar applications. *Appl. Opt.* **2017**, *56*, 2886. [CrossRef]
21. Javed, F.; Zhang, H.; Gao, Q.; Li, X.; Imran, A.; Jiang, Y. A high Average Power, Compact 100 kHz, 11.6-ns Nd:YAG Innoslab Amplifier. *Results Phys.* **2020**, *16*, 102926. [CrossRef]
22. Zhang, X.; Ye, J.; Luo, X.; Chen, X.; Zhang, L.; Xu, X.; Ren, H.; Lu, Y.; Ma, Y.; Gao, Q.; et al. 148-W single-frequency Nd:YAG Innoslab μs-amplifier at 1064 nm with high efficiency. *Opt. Express* **2021**, *29*, 5687–5693. [CrossRef] [PubMed]
23. Zhang, X.; Ye, J.; Luo, X.; Chen, X.; Zhang, L.; Xu, X.; Ren, H.; Lu, Y.; Ma, Y.; Gao, Q.; et al. 300 W two-stage Nd:YAG Innoslab microsecond amplifier. *Appl. Opt.* **2021**, *60*, 971–975. [CrossRef] [PubMed]

https://doi.org/10.3390/photonics13050449

MDPI AG
Grosspeteranlage 5
4052 Basel
Switzerland
Tel.: +41 61 683 77 34

Photonics Editorial Office
E-mail: photonics@mdpi.com
www.mdpi.com/journal/photonics

www.ingramcontent.com/pod-product-compliance
Lightning Source LLC
LaVergne TN
LVHW070444061126
851858LV00011B/158